30X

e d

The Red Ape

The Red Ape

*Orang-utans
and Human Origins*

JEFFREY H. SCHWARTZ

Houghton Mifflin Company / Boston
1987

Library of Congress Cataloging-in-Publication Data
Schwartz, Jeffrey H.
The red ape.
Bibliography: p.
Includes index.
1. Human evolution. 2. Orangutan. I. Title.
GN281.S33 1987   599.88′0438   86-20880
ISBN 0-395-38017-0

Printed in the United States of America

V 10 9 8 7 6 5 4 3 2 1

Diagrams on pages 62, 69, 79, 80, 84, 221,
224, 234, 239, 241, 243, 245, 250, 255, 258,
280, and 285 by Mary Reilly

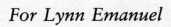

For Lynn Emanuel

Contents

Preface · ix

1 · The Orang-utan · 1

2 · In Pursuit of Human Ancestry · 26

3 · The Search for Our Closest Relatives · 57

4 · The Riddle of Relatedness · 86

5 · Primates First, Hominoids Last · 121

6 · The Hominoids · 156

7 · Humans and Orang-utans · 181

8 · Blood's Thicker Than Bones · 219

9 · Of Molecules and Evolution · 253

10 · The Return of the Red Ape · 283

Selected Bibliography · 309

Index · 317

Preface

IT'S FUNNY how things happen.

Twenty years ago I entered Columbia College thinking that I would be a psychiatrist. By the time I emerged from those four undergraduate years, I had been dissuaded of the notion of probing people's individual minds and infused with a desire to probe their collective past. I had taken a mandatory introductory course in cultural anthropology, with Robert Murphy, and then had to take one in physical anthropology, with Ralph Holloway — and by then it was all over: human evolution, here I come. Next I took a course on the human skeleton, followed by an independent study with Harry Shapiro, curator of physical anthropology at the American Museum of Natural History. Since then the dust of bones, fossils, and museums has been a part of my being.

Then on to graduate school — still at Columbia, still with Holloway. But things were getting hotter than ever in paleoanthropology. Everybody seemed to be at everyone else's throat. The Leakey family was splitting up. Holloway and Phillip Tobias were at odds about hominid brain evolution. Molecules were being pitted against teeth and bones. There seemed to be little room for new students of human evolution. And why get involved in all that heat, anyway?

A year of study at the University of London, University College, with Bob Martin provided me with a shift in focus. Bob introduced me to lemurs and lorises and fossil prosimians. At the same time I was "adopted" by Theya Molleson and Prue Napier at the British Museum (Natural History) and further introduced to modern primates, various analytic techniques, and a certain sense of history. All of these people gave me an education, on many levels, that could not have been predicted.

By then, however, the study of prosimian evolution was beginning to be as competitive as the study of hominids. I carved out a niche for myself, largely through the guidance of Theya Molleson, on the use of data on dental development and tooth eruption in sorting out evolutionary relationships. I chose work on the prosimians for my dissertation.

Back in New York City I got involved in the study of modern and fossil lemurs with Ian Tattersall, who had replaced Harry Shapiro at the American Museum. Through Ian I met Niles Eldredge, who was making quite a stir in theoretical quarters of evolutionary biology, and the two of them introduced me to a new way of looking at evolutionary problems, especially models of evolutionary change and methods of investigating evolutionary relatedness. Ian and Niles were not alone at the American Museum in thinking differently from most scientists about these issues, and I eventually came to understand the depth of their concerns as I listened to the sometimes screeching debate between opposing schools that ensued at the monthly systematic talks held there. I learned alternative ways of thinking about things, but I always had to account for myself with Walter Bock, a professor of evolutionary biology at Columbia. Let us say that Walter and Niles do not agree about certain aspects of evolution. But it was precisely in having access to such opposing viewpoints that I was most fortunate.

Being lucky is half the battle. I started with hominids, went into prosimian systematics, and now find myself back with human evolution. The faces may have changed, but the theoretical concerns have not. The study of hominids remains more volatile than the study of any lemurs could be.

One day about five years ago, while I was taking a break from research on some fossil lemurs at the British Museum (Natural History), Peter Andrews and two of his students brought up the subject of human and ape evolution. I had published a review article on primate evolution and had gone along with the commonly accepted notion of the unity of humans and apes, particularly humans and the African apes. Peter asked what I thought kept these primates grouped together evolutionarily. I had to admit that I really did not know. And that did it: I had to know why. Since that day, there have been some extremely important events of fossil discovery that have turned the study of humans and apes upside down. With these the orangutan has become an increasingly central figure in all major debates.

This book is my attempt to bring these arguments to a wider public.

In these pages I have tried to keep references to the literature accurate without letting them intrude upon the text — especially if they would interrupt the flow of an important argument. I have cited the relevant publications of mine in the bibliography. My articles in *Nature* and *Current Anthropology* contain some discussion of biomolecular and chromosomal data; the article in *Current Anthropology* and others present the information on the brain, handedness, and female hormone levels and additional theoretical problems. My discussion in this book of these topics represents more the history of my thinking about these issues than an account of my earlier publications. In the list of references, I have cited reviews whose own bibliographies deal quite adequately with historical information. Citations are made to specific papers I have discussed at length, except in a few instances, when one or two articles by the same author are summaries of, and make reference to, a series of his or her papers. The bulk of the research for this book was completed in October 1985, and therefore publications that have appeared since then may not be represented.

Many colleagues listened, corresponded, debated, infuriated, and otherwise contributed to my present state of perception about things broadly evolutionary or details more specific to the evolution of humans and apes; I thank them all. In particular, my thanks go to Peter Andrews, Wes Brown, Matt Cartmill, Desmond Clark, Nancy Czekala, Eric Delson, Niles Eldredge, John Fleagle, Steve Gaulin, Morris Goodman, Bill Jungers, Bill Kimbel, Bill Lasley, Jim Lennox, Roger Lewin, Jerry Lowenstein, Andrew Lumsden, Jon Marks, Lawrence Martin, Bob Martin, Henry McHenry, Mary McKitrick, Theya Molleson, Prue Napier, Bob Raikow, Peter Rodman, Vince Sarich, Susan Shideler, Charles Sibley, Elwyn Simons, Ian Tattersall, and Steve Ward. Their input was valuable, even though they might disagree with the way I processed the information. Thanks also to an anonymous evolutionary biologist who reviewed the completed manuscript.

Among my professional colleagues, many went out of their way to provide their original photographs or slides of primates for inclusion in this book. I thank Janice Crandlemire, Steve Gaulin, David Haring, Warren Kinzey, John Mitani, Herman Rijksen, Peter Rodman, Randy Susman, Russ Tuttle, and Pat Wright for their generosity and for giving me the opportunity to choose among the choicest of their photographic efforts.

A handful of individuals worked extremely hard to help me see this project to fruition. Alison Bond and Gerard Van der Leun saw through the awkwardness of my initial chapters to something they thought had merit. Through the months of writing and revision, Gerard stayed with me, pushing, prodding, and cajoling me away from the stiltedness of academic prose. His care and the additional attention to detail and continuity provided by Katya Rice, Bill Strachan, and Tania Wilcke provoked me to the end to rethink and rework ideas throughout the book.

Last, I want to thank John C. Anderton for his artistic contributions to the book. John took my notes on possible illustrations, seemingly memorized my manuscript, and then spent months working in the collections of the National Museum of Natural History and at the National Zoo to prepare the background sketches for the incredibly beautiful illustrations of live animals, skulls, teeth, bones, and combinations of these that are amply distributed throughout the text. Given his attention to detail, John more than once pointed out areas in my discussion that I needed to refine. But John's efforts were also important to me in that he drew what he saw in the specimens, not what I told him to draw from a specimen of my choosing. In most cases, I did not know until I received his finished illustrations in the mail if John had corroborated my own observations or found something different. His drawings are thus not just pleasurable to look at; they set a standard for accuracy and detail as well.

The Red Ape

1

The Orang-utan

THE HAND of a chimpanzee extends from the left lower edge of the photograph, its index finger outstretched. A woman's hand descends from the opposite corner. Her index finger points down toward the chimpanzee's. The tips of the two fingers do not quite touch.

Below the photograph is the logo of the Gulf Oil Corporation, the sponsor of a public television series that included a documentary on Jane Goodall's work with the chimpanzees. The photograph is a frame from that documentary. The woman's hand belongs to Jane Goodall. The chimpanzee's hand belongs to one of the animals Goodall has been studying as part of her almost twenty-year commitment to the study of wild chimpanzees. But I wonder if Michelangelo would have approved of the imagery, even if the chimpanzees had cooperated in assuming the poses of those souls, angels, and archangels who almost leap out at you in the painting of Creation in the shadow of the Sistine Chapel's ceiling, where the finger of God descends toward that of Adam.

The caption to the photograph asserts that the chimpanzee is our closest living relative, and indeed Jane Goodall has figured prominently in providing the general public with the images and movie film that make the idea of our sharing an intimacy with the chimpanzee quite appealing. In the longest continuous field study of the behavior of any nonhuman primate, Jane Goodall has certainly earned a place in primatological history. With each documentary and popular article the point is driven home: the gap between humans and chimpanzees is as narrow as it can be, short of the two actually touching.

Forays into the field to study the behavior and ecology of animals were not uncommon before Goodall began her work. Concerted, long-term studies were, however. Hardly anyone knew more than the old explorers' tales about the great apes — the chimpanzee, the gorilla, and the orang-utan. How could one get even an inkling of insight into human evolution if nothing of substance was known about our closest living relatives?

These concerns came to a head in the noted human paleontologist and archeologist Louis Leakey. Leakey had a personal investment in trying to unravel the mysteries of human evolution. He envisioned an intense research program of field studies on the three great apes: one investigator for each kind of ape, money for field support, and a commitment on each investigator's part to years of field research. Eventually all three apes were covered, with Jane Goodall concentrating on chimpanzees, Dian Fossey taking on the mountain gorilla, and Biruté Galdikas focusing on the orang-utan. (Of course there are, and have been, other students of the behavior and ecology of these great apes, but these women are the Leakey three and the ones popularized by the National Geographic Society.)

Especially because of Goodall's influence as one of the first to study chimpanzees seriously in the wild, and because of her academic visibility and her ability to generate public interest in the chimpanzee, this ape became the best known. Chimpanzees were studied in the field and in the lab, in natural and in controlled situations. Their humanness was constantly sought and emphasized.

Of the species of great ape, the chimpanzee is in many ways the easiest to study in the wild. This is not to say that, especially in the more thickly forested regions, chimpanzees are not at times difficult to follow and observe. It took years to track the so-called pygmy chimpanzee, which generally lives in denser vegetation than its larger sister species, the common chimpanzee. Still, the usually large group size and the overall gregariousness of chimpanzees make them more easily observable subjects than gorillas and — especially — orang-utans.

The forests of Africa are one thing, but the tropical rain forests of southeast Asia are something else — which is probably why the orangutan was the last of the large humanlike apes to be studied intensively in the wild. Colleagues of mine who do research in Hong Kong, Singapore, Djakarta, and other southeast Asian cities tell of having to bathe and change their clothes at least three times a day. But imagine being in such a climate in the pursuit of orang-utans, far away from the nearest city (or village, for that matter), in a place where a temporarily refreshing shower and change of clothing are confined to wishful thinking. Imagine being there with the British zoologist John MacKinnon, who composed the following in summary of months spent tracking the orang-utan, the red ape, on the southeast Asian islands of Borneo and Sumatra:

> Field conditions in the rain-forest were very severe. Steep terrain and thick vegetation made travel difficult. The dense canopy and the height of arboreal animals from the ground made observation conditions generally poor. As orang-utans were shy, quiet and dispersed they were hard to locate and easy to lose again. Heat, high humidity, rainstorms, floods and gales added to the discomfort and hazard of fieldwork. Leeches, wasps, mosquitos, horseflies and ticks added further problems and bears, wild pigs, snakes, crocodiles, elephants, banteng, and in Sumatra, tigers also produced anxious moments. Several hours of work in these conditions were required for each hour of observation achieved.

But what a sight it must be when you finally do get a glimpse of an orang-utan far up in the canopy of the forest: a scraggly, ruddy-haired, somewhat potbellied large ape, with incredibly long arms and double-jointed hips, probably strung out like a four-legged spider between the ends of four different branches of perhaps even different trees. It is hard to believe that an animal as large as the orang-utan, with males averaging about one hundred sixty-five pounds and females about eighty-two pounds, would be found so high in the trees and clambering about the ends of the finer branches, but that is usually where you will see them feeding.

Well over a century ago, Alfred Russel Wallace recorded his observations of orang-utans in his journeys through the Malay Archipelago. Better known for his having almost beat Charles Darwin to the punch in the publication of a theory of evolution based on natural selection, Wallace was also one of the most active natural historians of his day. His comments on the orang-utan's dietary behavior, written in 1869, are remarkably accurate — and note that in his dedication to the task he seems to have tasted some of the items in the diet:

> He feeds all through the middle of the day, but seldom returns to the same tree two days running . . . Their food consists almost exclusively of fruit, with occasionally leaves, buds, and young shoots. They seem to prefer unripe fruits, some of which were very sour, others intensely bitter, par-

FIGURE 1A   A "family" of Bornean orang-utans.

ticularly the large red, fleshy arillus . . . They almost always waste and destroy more than they eat, so that there is a continual rain of rejected portions below the tree they are feeding on. The Durian is an especial favorite.

Aside from the staples — fruit, leaves, buds, young shoots, and even bark — orang-utans also seek dietary supplements: epiphytes (plants, like orchids, that grow on other plants but are not parasitic on their hosts), lianas (probably best known as what Tarzan used for swinging through the trees), and wood pith. MacKinnon also observed orang-utans eating insects occasionally and suspected that they might sometimes eat small animals as well as birds' eggs. Orang-utans also seek out mineral-rich soils and eat handfuls at a time, much as a deer or a cow goes for a salt lick.

The orang-utan is certainly suited for traveling through the upper ches of the tropical rain forest canopy. Its long, curved fingers and toes serve as hooks, like the large claws of the sloth. The thumb and big toe are not right next to the other digits — rather, they ride high up on the hand and foot — and are markedly smaller. (In fact, the big toe is even sometimes absent.) The position and small size of the thumb and big toe increase the effectiveness of the longer digits. The overall length of the hand and foot is added to by elongation of the bones that sit just above the digits, the metacarpals of the hand and the metatarsals of the foot.

The orang-utan's arms are also very long. In both males and females, the arm is about twice as long as the distance from the rump of the animal to its neck, which is a distance close to the length of the animal's leg. Although the humerus, the upper arm bone, is long, much of the elongation of the arm comes in the two bones of the lower arm, the radius (whose lower end is on the thumb side of your arm) and the ulna (whose lower end is on the pinkie side).

In orang-utans, then, forelimb (arm) elongation, especially of the bones from the elbow on down, and a bit of streamlining, through diminution or even loss of the thumb or big toe, are characteristic. In contrast to the forelimbs, the legs appear short and atrophied. The general disposition of the arms and legs of the orang-utan was noted by a member of a group of British ambassadors, a certain Mr. M'Leod. He wrote about an orang-utan that had been captured in Borneo and subsequently accompanied his entourage back to England in 1816 after an unsuccessful mission to China: "His thighs and legs are short

and bandy, the ankle and heel like the human; but the fore part of
the foot is composed of toes, as long and as pliable as his fingers,
with a thumb a little situated before the inner ankle; this conformation
enabling him to hold equally fast with his feet as with his hands."

The orang-utan is also described in an earlier piece, *Telliamed; or,
Discourses Between an Indian Philosopher and a French Missionary
on the Diminution of the Sea, the Formation of the Earth, the Origin
of Men and Animals, Etc*. In this fantastic work, the early eighteenth-
century traveler Benoit de Maillet (whose last name spelled backward
is Telliamed) wrote about an expedition in 1702 of the Dutch East
India Company from the city of Batavia (now called Djakarta) on the
island of Java, Indonesia, to the island of New Guinea, near Australia.
It was during this expedition ("which was of no use") that the Dutch
captured two male animals, "which, in the language of the country
where they were taken, [were] called *Orangs Outangs*." These "Orangs Outangs" were characterized as having thick hair, long and
somewhat crooked fingers and toes, and feet that "were flat where
they are joined to the leg . . . so that they resembled a piece of plank
with a baton driven into it."

We may be struck initially by the great length of the orang-utan's
arms, but perhaps even more astonishing is the unique ability of the
orang-utan to move its legs in arcs as easily as it (or you and I, for
that matter) can move the arms. Indeed, the orang-utan's leg is essentially and functionally an arm, with fingerlike toes and a shoul-
derlike hip joint.

A feature of primates in general — and we, apes, monkeys, and
lemurs and lorises constitute the group called Primates — is the great
mobility of the shoulder joint. It is thought that there is a correlation
between the degree of rotation around the shoulder joint and the
degree to which the animal is arboreal, or spends time in trees. Our
greatly mobile shoulder joints are supposed to reflect our arm-swinging, tree-living heritage. On the other hand, our hip joints, and the
hip joints of most primates, are securely strapped into place by a
ligament that holds the ball of the femur, the thigh bone, into the hip
socket of the pelvis. This ligament, however, is absent in the orang-
utan. There is nothing to restrain the movement and posturing of an
orang-utan's leg as it dangles itself in whatever position is necessary
to pick its way through the trees and feed on the offerings at the ends
of the finest branches.

But what is good in the trees may not be so good on the ground. The orang-utan, although typically not fast-moving in the trees, is at least comfortably acrobatic there. On the ground, however, an orang-utan is slow and awkward. Its arms seem particularly inappropriate and its long fingers and toes a nuisance. It shuffles along on the outer edges of its hands and feet, its fingers and toes being flexed or cupped so that there appear to be four weakly formed fists.

Mr. M'Leod, who took a bit of poetic license in describing the typically potbellied orang-utan as "looking like one of those figures of Bacchus often seen riding on casks," commented in a less fanciful vein that the orang-utan's "natural locomotion, when on a plain surface, is supporting himself along, at every step, by placing the knuckles of his hand upon the ground." But although the knuckles of the hand and foot may touch the ground while the orang-utan is moving along, its mode of getting around on the ground should not be confused with what, in the chimpanzee and the gorilla, is called "knuckle walking."

On the ground the fingers and toes of an orang-utan are typically tucked under the hands and feet. In contrast, the foot of a chimpanzee

FIGURE IB   An orang-utan *(left)* and a gorilla *(right)* in typical quadrupedal stance. The gorilla, a knuckle walker like the chimpanzee, plants its feet flat on the ground and supports its forward body weight by resting on the surfaces of its bent fingers. The orang-utan shuffles along on the sides of its hands and feet, the elongate digits of which are loosely cupped.

FIGURE 1C   A typical Sumatran *(left)* and a typical Bornean *(right)* adult male orang-utan. As is characteristic of fully mature adult males, both have well-developed cheek pads, throat sacs, beards, and mustaches. However, their faces look strikingly different.

or a gorilla when on the ground is planted flat on the surface, just as our own foot is, and the animal's hand is held in line with the straightened arm and its fingers are flexed at the middle knuckle. The weight of a chimpanzee or a gorilla is placed on the knuckles of all four fingers. As has been well known for over a century, the muscular anatomy of chimpanzee and gorilla arms is special and seemingly correlated with the posturing and the knuckle walking of these African apes.

Not all orang-utans look the same. There are, for example, differences between Bornean and Sumatran orang-utans in the shape and configuration of the face. As John MacKinnon pointed out, the face of the Sumatran orang-utan is long and oval, whereas the face of the Bornean orang-utan is thicker and broader. Viewed side by side, the two are strikingly different. Mr. M'Leod, in reference to his Bornean orang-utan, commented that "the lower part of his face is what may be termed an ugly or caricature likeness of the human countenance." Another early traveler described the head of a Sumatran orang-utan as being "well proportioned to the body."

The differences between the two are even more exaggerated in mature adult males because of their development of "cheek flanges" or "cheek pads," massive thickenings of the cheek region. They are not present in newborn male orang-utans, and they rarely develop in female orang-utans. Cheek pads are noticed first in young male orang-utans as a toughening of the skin on each cheek and can then develop into extensions from the side of the forehead, above the eyes, to below the edge of the lower jaw. The cheek pads of the Bornean male orang-utan are quite heavy and large, often rather floppy, and somewhat square. Those of the Sumatran male orang-utan are slimmer and less pronounced and have been described by MacKinnon as somewhat diamond-shaped in appearance. It is perhaps not surprising that the face of a Sumatran orang-utan could be described as being better "proportioned" than the face of a Bornean orang-utan.

Bornean and Sumatran orang-utans also differ in their facial hair. Both types, especially the males, develop facial hair that can be identified as a beard and a mustache, but the beard and mustache of the Sumatran male orang-utan are much nicer looking and fuller. And while the cheek pads of a Sumatran male orang-utan are covered with tufts of feathery hair, the Bornean male has short, bristly, widely separated hairs protruding form his dark, lumpy cheek pads.

Although females may develop throat sacs, fully mature males are distinguished by the massiveness of theirs. In some animals the throat sac serves as a vocal resonating chamber. In orang-utans, however, the sac is just a large, loose fold of skin hanging from the neck. The flap makes females look bloated to one degree or another beneath the lower jaw, but in mature males, the sac is so extensive that it flows down from behind the lower edges of the cheek pads. In short, a mature male orang-utan looks a lot like a big, saggily goitered, hairy, reddish Buddha wearing a baseball catcher's mask.

Among the Sumatran orang-utans, one can distinguish two distinct types. According to Herman Rijksen, a Dutch scholar who has become a leading authority on the Sumatran orang-utans, one of the major features that readily distinguishes one type from the other is the color of the hair. One kind of Sumatran orang-utan has dark brown to maroon hair. The other variety has lighter hair, ranging in color from reddish cinnamon to rusty red.

While reading one of the many editions of *Buffon's Natural History,* a popular eighteenth- and nineteenth-century collection of zoological fact and folklore, I came across a brief description of a Sumatran orang-utan that had been shot at Ramboon, near Touraman. This is on the northwest coast of the island, about two days' journey from the dense forest in which orang-utans are normally found. The emphasis of the passage was on the extremely large size of the specimen, which was supposed to have been between seven and eight feet tall, and would therefore have been much taller than any Sumatran (or Bornean, for that matter) orang-utan ever captured or observed. For almost a century after the description was published in 1826, by a Dr. Clarke, the notion of a "race" of gigantic Sumatran orang-utans continued to pervade the literature. Only in 1912 did the American primatologist D. G. Elliot question the accuracy of the reported size of the specimen. Elliot suspected that the measurements had been taken from a skin that had been greatly stretched and distorted. He also pointed out that "if there was a race of gigantic Ourangs in north west Sumatra, it would be strange that during the last century another example had not been obtained."

In *The Malay Archipelago*, Alfred Russel Wallace recorded some measurements of the height and arm span of "the bodies of seventeen freshly killed Orangs," of which "sixteen were fully adult, nine being males and seven females." Measuring each specimen "fairly to the heel, so as to give the height of the animal if it stood perfectly erect," Wallace found that "the adult males . . . only varied from 4 feet 1 inch to 4 feet 2 inches." Measuring between the fingertips of the outstretched arms of the male orang-utans, Wallace found the span to vary between seven feet two inches and seven feet eight inches.

The only specimen of a female orang-utan for which Wallace reported measurements stood three feet six inches tall and had an arm span of six feet six inches. The height of this specimen is approximately eighty-four percent of the height of the males Wallace measured, which is precisely what the British zoologist Colin Groves (now living in Australia) found to be the difference between males and females in his more extensive metric study of the physical dimensions of orang-utans.

Apart from the matter of the animal's size, the early nineteenth-century description of the "giant" Sumatran orang-utan is accurate for features of a Sumatran orang-utan. The skin is described as being "of a dark leaden colour, covered, unequally, with brownish red, shaggy and glossy hair, which is long on the flanks and shoulders." Given the description of its beard — "The chin was fringed with a curling beard, reaching from ear to ear" — it is likely that this specimen was male.

In writing about the orang-utan of Borneo, Wallace typically used the native Dyak name "Mias" instead of the Malay word "orang." In Dyak, a mature adult male orang-utan with fully developed cheek pads is called "Mias Chappan" or "Mias Pappan." The Dyaks use "Orang" strictly to refer to a man, which is what the word means in Malay.

According to the Dyaks, the Mias is rarely attacked by other animals of the forest, and when it is, only by the crocodile and the python. One old Dyak chief interviewed by Wallace declared that "no animal is strong enough to hurt the Mias." In a fight between an orang-utan and a crocodile, "the Mias gets upon him and beats him with his hands and feet, and tears him and kills him."

The strength of the orang-utan is legendary. A female orang-utan is over three times stronger than a well-conditioned young human male. A male orang-utan's strength is more than four times as great as his human counterpart. In tests where a dynamometer was used to assess the strength of the grip of orang-utans, a male orang-utan squeezed the device with fifty percent more strength than did a female weighing about four-fifths as much. No wonder one of Wallace's informant chiefs, Orang Kaya, claimed that the Mias could kill a crocodile "by main strength, by standing upon it, pulling open its jaws and ripping up its throat."

Given the brute strength of the orang-utan, as well as its bulk (even of the female), it is astonishing to think of the animal swinging along perhaps thirty or forty feet up. Granted, as Wallace noted, the orang-utan "never jumps or springs" — and perhaps wisely. Instead, the orang-utan gets from tree to tree by carefully choosing the seemingly safest of intermingling branches and then cautiously but deliberately swinging across to the next tree.

In the course of getting about or foraging for food, the orang-utan does quite a bit of damage, sometimes breaking off entire branches, which drop to the forest floor with a resounding crash. Wallace noted, too, that the orang-utans "almost always waste and destroy more than they eat, so that there is a continual rain of rejected portions below the tree they are feeding on." When night approaches, an orang-utan typically makes a sleeping nest from the leaves and branches of the tree it has chosen. A shower or sudden downpour will cause the orang-utan to cover itself with large leaves or ferns, constructing a "rain hut."

Biruté Galdikas has argued that the apparently destructive nature of the orang-utan is actually beneficial to the survival of the animal's own forest. For example, by eating buds, young shoots, and even unripe fruit, the orang-utan is essentially pruning the tree it is feeding in. By consuming whole fruits, orang-utans, like birds, become unwitting agents in the dispersal of plant seeds. Finally, because the orang-utan tries to support itself on slim branches, branches are broken, thus letting light into the lower reaches of the dense forest. Thinning of the forest canopy also results from the orang-utan's construction of nests and rain huts.

After observing orang-utans for some time, Wallace commented that he "never saw two full-grown animals together," and indeed there has been a tendency through the years to characterize the orangutan as a solitary animal with no definable social organization. The impression of the unsociability of orang-utans is, however, being altered through the long-term studies of such field researchers as Rijksen, Chris Schürmann, MacKinnon, Peter Rodman, John Mitani, and Galdikas. Nevertheless, the fact remains that orang-utans are widely dispersed. When you consider how big and heavy an orang-utan — even a female orang-utan — is and that much of the time an orangutan spends feeding is at the thin ends of branches, it makes sense that individuals of the same social group will space themselves throughout the forest. Having individuals widely separated is also consistent with the fact that the foods an orang-utan typically seeks are not usually located or concentrated in just one tree.

Trees, and plants generally, do not produce their buds, shoots, or fruit all at once. Especially in tropical rain forests, where the climate is fairly consistent and nonseasonal, the period over which a tree will go through the various phases of its growth and reproduction can be quite long. Animals in search of certain foods — a particular growth stage of a particular leaf or fruit — must range through at least a few trees to ferret out the right dietary bits and pieces. The larger the animal, the more it will need to consume, the more trees it will need to visit in a bout of foraging, and thus the more territory it must define as its own. MacKinnon estimated that in Borneo there are only one or two orang-utans in just over one square kilometer of forest. For Sumatra, MacKinnon concluded that each orang-utan needed an even larger area in which to forage.

Even with this extreme spatial isolation of individuals, however, there is method to the movements of orang-utans. Orang-utans are organized loosely as a group around an adult male. Individual orang-utans may not be in visual contact with each other, but group cohesion is maintained by very long, loud calls emitted by the adult male, which keep the others of his group, as well as possible trespassers, aware of his general location. If the sound gets too distant, you know you should move to catch up with it. If the call is loud and you don't belong to that male's group, you had better distance yourself immediately. In addition to its function in the spacing of group members and in keeping others away, the male's long call also appears to play an important role in mating and courtship behavior.

Orang-utans do not act like solitary animals when it comes to mating. Really solitary primates, such as the nocturnal bushbabies of sub-Saharan Africa, mate only when a male and female whose defended territories overlap bump into each other during their nighttime foraging *and* when the female is in heat (estrus). It is only during estrus that most mammals, including most primates, can mate. At any other time of the menstrual cycle, mating for the female is physiologically and physically impossible — even if the opportunity to copulate were to present itself. The period of estrus is timed to the menstrual or ovulatory cycle: a female comes into estrus at the peak of fertility, which is during ovulation. And there are physical signals, such as the ballooning up of the genital region, that coincide with peak fertility. Such signals alert males to the female's sexual receptivity and thus to the opportunity to produce offspring of their own if their mating is successful. Among animal behaviorists, estrus is defined as having two components: behavioral (the restriction of mating to a certain part of the menstrual cycle) and physical (at least some genital swelling).

Female orang-utans are unique among the apes in that they do not have an estrous cycle, with its behavioral constraints and external signs, imposed upon their menstrual cycle. As demonstrated in laboratory studies and observations on zoo animals, female orang-utans, given the chance, copulate throughout the menstrual cycle. Orang-utans in the wild copulate more frequently near the peak days of fertility but are not limited to that time. In captive chimpanzees and gorillas, as in wild chimpanzees and gorillas, copulation is restricted to the period roughly around ovulation.

As ovulation approaches, the genital region of a female chimpanzee becomes greatly engorged. The visibly unavoidable swelling attracts the attention of many males during her estrus. Gorillas also have some physical signs of estrus, but the swelling of the female's genital region is, at least to human observers, not as blatant. Female orang-utans in both captive and natural settings have not been found to show any physical changes in the genital region at ovulation or any other phase of the menstrual cycle. Curiously, and in contrast to all other apes and to primates in general, a female orang-utan will develop swelling of the genital region *after* she has become pregnant.

Adult male and female orang-utans form consortships that last for at least a few weeks. During that time, they travel together and copulate frequently. It is during this period, and only during this time, and only by her consort, that a female orang-utan becomes pregnant. Now that field studies of the same forested areas and the same orangutans have been carried out over periods of years, it has become apparent that the same male and female continue to form consort pairs for more than one mating period. Although there may be an occasional sexual attack on a female by an overly aggressive young male, such attacks are reproductively unsuccessful. By forming a consortship, a partnership, with an adult female for most of her monthly cycle, an adult male orang-utan assures himself of his parenthood, removing any doubt that the offspring he might eventually see with her is his offspring. Returning to, and being accepted by, the same female when she is ready to become pregnant again — and it can be as long as seven years between orang-utan births — adds, I am sure, another level of social and parental certainty. Certainly this is a far cry from the situation with the chimpanzee, which has been classified by sociobiologists as promiscuous. All hell breaks loose when a female chimpanzee comes into estrus, and when it is over nobody knows who the offspring's father was.

Perhaps because their lifestyle demands long periods of isolation and perhaps because they don't develop sexual swellings that serve as enticement, female orang-utans take an active role in soliciting sexual advances and attention from the male. This happens not just at the beginning, when a female might be trying to initiate consortship, but during the entire period of partnership. Especially on the part of the female, there is a lot of touching of the partner with the fingers and the lips before and during the couple's very long copulatory bouts. In the wild, these acts are even more intense and prolonged. Rather than mounting the female from behind, as a dog does, for example, the male orang-utan frequently mates in a face-to-face position with the reclining female. There are occasional reports of this position being used by chimpanzees, but the male chimpanzee tends to mount the female in the more typical mammalian fashion. The active role of the female orang-utan in mating also includes her helping to guide the male's penis to successful intromission.

An obvious comparison of the orang-utan, at least in its sexual and reproductive behavior and physiology, is to humans. Human females do not develop sexual swellings at any time during the menstrual cycle, copulation is possible at any time during the menstrual cycle, and copulatory bouts are quite long in contrast to the seconds it takes most other primates, including the chimpanzee and gorilla. The frequency with which human females copulate increases when ovulation approaches, as for orang-utans in the wild, but there are no physiological, physical, or behavioral barriers to copulation at times other than during the brief peak period of fertility. Taking an active sexual role is also true for human females.

It might seem strange to compare humans and orang-utans — especially when we have virtually been instructed to believe that the African apes, and probably more the chimpanzee than the gorilla, are our closest living relatives. But if history and received wisdom had held the weight that they usually do in preventing old ideas from being replaced by the new, then we would all have learned first about orang-utans, and then about the other apes. It was the orang-utan that first captured the attention of natural historians and adventurers as the most humanlike of the primates.

Well before the earliest threads of evolutionary ideas appeared in print, there was a view of the world that, superficially at least, resembled evolution. This world view proposed a hierarchical ordering of organisms from the lowest to the highest. It was an old idea, dating back as far as the fourth century B.C., when the Greek philosopher, anatomist, and natural historian Aristotle introduced, according to later historians and philosophers, the notion of a "ladder of life" or "ladder of creation." As it was interpreted, Aristotle's concept likened a sequence of organisms to the rungs of a ladder, with each lower form striving, or at least wishing, to attain a higher rung. Of course, humans, the gods' highest form of creation, occupied the most illustrious rung.

The idea of a hierarchical arrangement of things on earth lasted into the next millennium, surviving many centuries that were devoid of active thought and intellectual inquisitiveness. In the seventeenth and especially the eighteenth centuries there was intense renewed interest in filling in the gaps in the ladder of life, this natural hierarchical ordering of all things. The task was in no small way made more possible by the concurrent explosion of interest in exploring foreign and mysterious lands and in examining the rocks, minerals, plants, and animals of these new frontiers. Aristotle's ladder became characterized as the "great chain of being." And while the lower ranks were being filled in with all sorts of bizarre organic and inorganic "forms" (real or unreal, it sometimes didn't seem to matter), the link between monkeys and humans was also being sought. This position was to be filled by a conglomeration of fact and fantasy — a mythical beast called "Orang Outang."

In the older English literature, orang-utan is often spelled Orang Outang, but there are some understandable variants. Basically, however, "Orang Outang" and its linguistic relatives are all pronounced the same way (pretty much as you see it) and are Malay words meaning "the man of the forest." Malay is not spoken in Borneo but is found throughout northern Sumatra. Since Bornean and Sumatran orang-utans look more like each other than anything else, it is reasonable that the two would be called by the same name, but what happened by the latter part of the seventeenth century was that "Orang Outang" came to include the African apes, the chimpanzee and the gorilla, as well. All sorts of strange tales and images of "Orang Outang" emerged, making it quite difficult at times to figure out which animal — the real orang-utan or one of the African apes — was being discussed. But if we stay with the more decipherable sources, it is possible to unravel the picture.

The earliest images of the "Orang Outang" are of the real southeast Asian orang-utan. A perception of the humanness of this animal, setting it apart from all other animals, was the attribute most fequently emphasized. One early source was a seventeenth-century Dutch physician who wrote under the Latinized name of Bontius; he compiled a natural history of "India" (which at that time was under Dutch control and extended from what is now recognized as east India throughout much of southeast Asia). Bontius was stationed in Batavia, now Djakarta, on the island of Java, which is one of the three large islands of the southeast Asian archipelago that also includes the nearby islands of Borneo and Sumatra. James Burnett, also known as Lord

Monboddo, who wrote in the eighteenth century on the origin of language and for whom the Orang Outang played a very central role, quoted from Bontius in the Latin and then provided a paraphrased translation:

> He relates, that he saw several Orang Outangs, of both sexes, walking erect; and he particularly observed the female, that she shewed signs of modesty, by hiding herself from men whom she did not know. And he adds, that she wept and groaned, and performed other human actions: So that nothing human seemed to be wanting in her, except speech.

Benoit de Maillet wrote in the early eighteenth century in *Telliamed* of the true orang-utan that "they had the whole of the human form, and like us walked upon two legs." As it was popular to impart a quality of sensitivity that set orang-utans apart from the rest of the animal world, de Maillet described the captive orang-utans as being "very melancholy, gentle, and peaceable." As orang-utans were distinctly different from other animals and would be perfectly human if they were to master speech and language, de Maillet found it necessary to add that these orang-utans "could only articulate sounds very indistinctly."

Travelers' reports of chimpanzees and gorillas also began to emerge in the early 1700s, especially from forays into Angola. The gorilla was called "Impungu" by the natives of the region. And "Impungu" was no doubt the basis of the name "Pongo," which British explorers used frequently to refer to the gorilla. Interestingly, Pongo has since become the accepted genus name for the orang-utan — its taxonomic genus and species names are *Pongo pygmaeus*.

An anonymous merchant, who had had a trading ship that dealt in slaves on the west coast of Africa, wrote that there were three species of "Orang Outang" living in and around Angola. Of the gorilla, or Impungu, he noted that

> This wonderful and frightful production of nature walks upright like man; is from 7 to 9 feet high, when at maturity, thick in proportion, and amazingly strong; covered with longish hair, jet black over the body, but longer on the head; the face more like the human than the Chimpenza, but the complexion black; and has no tail.

All it takes is one good "story" and a major myth can take hold. The picture we are left with from tales like this is that everyone is afraid of the Impungu. And there is, according to the merchant, good reason for being afraid of the Impungu, since this "fierce" animal is supposed to make a habit of pursuing and capturing natives. As the story goes, if the Impungu does catch a native, it either kills him or takes him away as a prisoner. But occasionally a prisoner escapes; he waits, the merchant tells us, until the Impungu falls asleep, and then "steals away his hand or arm softly from his, and so steals away quietly." Sometimes, however, this freedom is short-lived and the prisoner is "discovered and retaken."

The chimpanzee was referred to as "Quimpezes," "Enjocko," "Jocko," and "Chimpenza." "Chimpenza" is the native African word for the chimpanzee. The Chimpenza was reported by the merchant to be the smallest of the three African "Orang Outangs," similar to the other two in overall appearance and shape but supposedly walking on all fours more frequently than the others. The story was that "the females have their times like women" and that male chimpanzees sought native females; when a male caught one, "it commonly forces, and lies with her." Chimpenzas were also credited with having "games and pastimes like the natives" and even with having a king who "does not work himself, but orders." Chimpenzas use fire, which they steal from the natives, "associate in communities, and build little towns or villages," which they leave after sleeping one night and rebuild elsewhere the next night.

The third African Orang Outang was called the "Itsena" and was described as being "less than the Impungu, and larger than the Chimpenza." The Itsena is "like the Chimpenza in every respect, unless in size," but "they keep to themselves, the Chimpenza and they not agreeing."

We know that the Itsena was an ape and not a monkey because the merchant's letter ended with a postscript stating that "all the three species have no tail." Apes and humans lack a tail, but monkeys retain the primitive mammalian condition of developing a tail. However, there is no known species of African ape that is intermediate in size between the gorilla and the chimpanzee. There is a second chimpanzee species, the pygmy chimpanzee, but it is a bit smaller than the common chimpanzee. And there are two subspecies of gorilla, but although one subspecies is slightly smaller than the other, the smaller of the two is nonetheless still huge. Perhaps the most reasonable interpretation is that the Itsena was a female gorilla.

Sexual dimorphism — that is, marked differences between the sexes in such features as size and shape and color — is especially dramatic in the orang-utan, in which the male can be much larger than and close to twice the weight of the female. Dimorphism characterizes the chimpanzee and the gorilla also, but while a male chimpanzee may be somewhat larger than a female chimpanzee, the degree of sexual dimorphism typical of the gorilla is closer to the extreme seen in the orang-utan.

A female gorilla typically lacks the massive bony buildup that enlarges the outward appearance of the skull and gives it the upward and backward bulge and the robust brow ridges so characteristic of the male gorilla. In its more smoothly oblong skull and less protuberant brow ridges and also in its smaller size, a female gorilla more closely approximates the general semblance of the chimpanzee. The confusion that marked sexual dimorphism can cause is apparent even in scholarly works of the early twentieth century, in which the skulls of various female gorillas are illustrated as representing not just different species, but different genera (the plural of genus). One such genus was *"Pseudogorilla,"* because the skull of the specimen looked more like a gorilla than anything else, but it didn't have all the hallmarks of a true, full-blown King Kong.

The conglomerate Orang Outang — which subsumed the real orang-utan, the Impungu or gorilla, the chimpanzee or Chimpenza, and the Itsena — was considered to represent, in some way, the missing link between monkeys and humans in the great chain of being. This intermediacy was supposed to have been demonstrated, for example, in 1699, with the publication of an oft-cited anatomical study by the British comparative anatomist Edward Tyson. The title of Tyson's treatise was *Orang-outang, sive Homo sylvestris; or, The Anatomy of a Pygmie Compared with That of a Monkey, an Ape, and a Man.* As far as Tyson was concerned, the anatomical similarities between the Orang Outang and humans were incontrovertible. In one of the lithographs in the volume, the "Orang-Outang" is even depicted standing upright, holding a staff.

The problems with Tyson's work, however, are many, including the fact, as the Harvard evolutionary biologist Stephen Jay Gould has pointed out, that the "Orang-outang" is not an orang-utan but a chimpanzee. There is also the fact that the specimen was a juvenile individual, which would not yet have developed the sloping forehead, the brow ridges, and the protruding jaws that would have made it look more chimpanzeelike than human. Futhermore, as Gould argues, Tyson was predisposed to filling the gap between humans and monkeys before he even undertook his study, so it is likely that "Orang-outang" would have been found to be the missing link anatomically no matter which ape of whatever age was studied.

The association of the Orang Outang with humans became more than just a matter of linking the "highest forms of creation" with the lower monkeys. It appears to have been a way of extending the hierarchy into the human species, to create subcategories of humans. Carl von Linné, the Swedish botanist who developed the organized scheme of classifying plants and animals that we still use today (and, by classifying himself along Latin lines, came to be known as Linnaeus), distinguished the "wild man" from other varieties of *Homo sapiens* on the basis of speech. The wild man Linnaeus defined as "four-footed, mute, and hairy." Below the wild man was the Orang Outang, which Linnaeus formally recognized as *Homo sylvestris,* the "man of the forest or of the trees." *Homo sylvestris* was known to be different from monkeys in lacking a tail and in having emotions. The wild man, which was based essentially on mythology rather than observation; was really no different from *Homo sylvestris* except that it perhaps looked more like other *Homo sapiens.*

The other varieties of *Homo sapiens* that Linnaeus defined were the "American," the "Asiatic," the "African," and the "European." Divisions within *Homo sapiens* were distinguished in general by their "varying by education and situation." The wild man was on the low end of the ladder. The European, at the highest end of the scale, was described as "fair, sanguine, brawny . . . hair yellow, brown, flowing . . . eyes blue . . . gentle, acute, inventive . . . covered with close vestments . . . governed by laws." It does not take a genius to figure out that the "intermediate" varieties of "man," although elevated above the wild man, were arranged at the base of the sequence leading to the European.

Among others who produced classifications of "man," one, Bory St. Vincent, created fifteen species, each of which was subdivided into a number of varieties. Bory St. Vincent also believed that each species of "man" had had "its peculiar Adam, and its peculiar place of origin." His first, or most elevated, species was the "Japetic," which was divided into four races, with the German race having two varieties. The sequence then descended through, for example, the "Arabian" (number two), the "Hindoo" (number three), the "Sinic" or Chinese (number five), the "American" (number ten), and the "Hottentot" (number fifteen and last on the list). And it was the Hottentot that Bory St. Vincent, like others, considered to be the link between humans and the Orang Outang.

The image of the Orang Outang as a "mute" link between the "lower" animals and eventually true or civilized man was an obsession of many scholars of the seventeenth and eighteenth centuries, not all of them taxonomists or natural historians. One obsessed scholar was James Burnett, Lord Monboddo, a Scottish philosopher who made himself well known not only through his treatises but through his taking on the likes of the equally vocal David Hume and Samuel Johnson.

One of the major lines of inquiry during the eighteenth century was the origin of language, and Lord Monboddo produced six volumes on the subject (published anonymously). The first volume of this twenty-year ordeal appeared in 1773 and dwelt at length on the Orang Outang. As far as Lord Monboddo was concerned, the Orang Outang was a "man," but one without a spoken language:

> I still maintain, that his being possessed of the capacity of acquiring it, by having both the human intelligence and the organs of pronunciation, joined to the dispositions and affections of his mind, mild, gentle, and humane, is sufficient to denominate him a man. And it is very extraordinary to suppose that he is of another species, not because he wants any organs that we have, such as the organs of speech, but because he does not make the same use of them.

Lord Monboddo proceeds to argue that the reason Orang Outangs do not use their "organs of speech" is that they do not live "long together in close intercourse" nor do they practice the arts. Thus, "it is not to be wondered, that, if men have had no occasion to live together in that kind of strict society, but have been able to subsist upon the natural fruits of the earth, with few or no arts, which is the case of the Orang Outang, they should not have acquired language."

The notion that the Orang Outang was a human trapped in the body of a mute, hairy beast, or might even represent primordial "man" himself, attracted the attention of the English satirist Thomas Love Peacock, who in 1818 published a two-volume work entitled *Mel-*

incourt; or, Sir Oran Haut-ton. "Haut-ton" can be translated loosely from the French as "highly mannered." Although Peacock took swipes at virtually every social contrivance of the day, the major subject of *Melincourt* was an Orang Outang who not only was accepted but became an important figure in local high society.

Peacock's Sir Oran Haut-ton is the conglomerate Orang Outang. The reader is told that "Sir Oran Haut-ton was caught very young in the woods of Angola," and is "called in the language of the more civilized and sophisticated native of Angola, *Pongo.*" This bit of information probably refers to the gorilla. However, reference is made to the fact that the animal is also called "Oran Outang" by the "Indians of South America," which has to be an error, and should instead read the "Indians of southeast Asia."

In one particular passage, spoken by the character Mr. Forester, Peacock summarizes the debate over the "status" of the Orang Outang. Lord Monboddo, whose position is that the Orang Outang is truly a man, is pitted against the likes of the powerful French philosopher Rousseau, who denies the Orang Outang a position on the same rung of the ladder as we:

> Some presumptuous naturalists have refused his species the honours of humanity; but the most enlightened and illustrious philosophers agree in considering him in his true light as the natural and original man. One French philosopher, indeed, has been guilty of inaccuracy, in considering him as a degenerated man; degenerated he cannot be; as his prodigious physical strength, his uninterrupted health, and his amiable simplicity of manners demonstrate. He is, as I have said, a specimen of the natural and original man — a genuine fac-simile of the philosophical Adam.

Sir Oran Haut-ton falls in love with a woman named Anthelia, but because he is not a "real man," he must content himself with only being near her. Thus, Sir Oran Haut-ton comes to "reside with Mr Forester and Anthelia," who eventually get married: "His greatest happiness was in listening to the music of her harp and voice: in the absence of which he solaced himself, as usual, with his flute and French horn. He became likewise a proficient in drawing; but what progress he made in the art of speech we have not been able to ascertain."

The debate over the origin of language, and the conditions in which truly modern "man" would arise, reminds me of Darwin's arguments for human origins in his *Descent of Man,* published in 1871, twelve years after *On the Origin of Species.* It was Darwin who first suggested that there was an evolutionary link between the apes of Africa and the human species.

By the time Darwin came to write *The Descent of Man,* the conglomerate Orang Outang was, for all intents and purposes, defunct. The separate identities of the chimpanzee, the gorilla, and the real orang-utan were recognized and the individual animals correctly referred to in the scientific literature. The major evolutionary work that had appeared on the relationships of humans to other primates was by Darwin's public defender, Thomas Huxley. In one of three essays published in 1863, Huxley mustered an impressive array of anatomical information to demonstrate that humans were not only primates, but most similar to the three great apes, the gorilla, the chimpanzee, and the orang-utan. Huxley, however, kept the three large apes together as a group, finding them more similar to one another than to humans. The conglomerate Orang Outang lived, but in a different form.

Darwin's case for the "descent of man" rested upon a preconceived notion of an evolutionary continuum. He held firmly to the idea that evolution was a gradual process and thus that the origin of new species was the result of accumulation of minor changes, generation by generation, over long periods of time. Missing links were predicted by Darwinian theory and were sought with as much conviction as they were by believers in the great chain of being. Since humans are a fleet and bipedal (walking on two feet) species, Darwin believed that there would be evolutionary evidence for the development of this condition:

> If the gorilla and a few allied forms had become extinct, it might have been argued, with great force and apparent truth, that an animal could not have been gradually converted from a quadruped into a biped, as well as the individuals in an intermediated condition would have been miserably ill-fitted for progression. But we know (and this is worthy of reflection) that the anthropomorphous apes are now actually in an intermediate condition; and no one doubts that they are on the whole well adapted for their conditions of life.

The arms of the great apes are typically longer relative to their legs than in many monkeys, especially the more commonly known and terrestrial types like the baboons. Because of its long arms, an ape appears to be holding its torso in a more erect position. If you are seeking intermediate forms and missing links, you might see apes as representing something intermediate between a quadrupedal and a fully bipedal animal, and then you can argue that such an intermediate animal is well adapted, as is obvious from the fact that it exists.

Darwin was not content to stop at the level of the argument of the great apes representing, in general, an intermediate anatomical phase in the "descent of man." Indeed, a major thrust of *The Descent of Man* was the evidence for the development of civilized humans from a much more savage and barbaric state, to demonstrate "that man has risen, though by slow and uninterrupted steps, from a lowly condition to the highest standard as yet attained by him in knowledge, morals and religion." But before Darwin could entertain the rise of "civilized nations," he had to argue first for the "birthplace of man."

Given that Darwin was the author of a theory of evolution based on natural selection and adaptation, it is no wonder that he used such arguments in developing a background for human origins. His ideas on the "descent of man," from its earliest evolutionary beginnings to its attainment of the "highest standard," were steeped in the notion of competition between social groups or tribes. But the major impetus for the whole process, in his view, was another level of competition, that of living in a dangerous environment.

Where would one look for the "birthplace of man"? Obviously, as Darwin concludes in chapter 2 of *The Descent,* one should not seek human origins in southeast Asia because potential "ancestors would not have been exposed to any special danger, even if far more helpless and defenseless than any existing savages, had they inhabited some warm continent or large island, such as Australia, New Guinea, or Borneo, which is now the home of the orang." But, for example, "no country in the world abounds in a greater degree with dangerous beasts than Southern Africa." Almost one hundred pages later, with an unfounded authority, Darwin asserts that as the chimpanzee and the gorilla "are now man's nearest allies, it is somewhat more probable that our early progenitors lived on the African continent than elsewhere." The conclusion rested not upon comparative anatomy but upon the observation that Africa was home to some apes as well as to supposedly primitive humans.

Most of the paleontological work related to human origins in the past century has been based on Darwin's conviction that the birthplace of the human species is in Africa, and this is itself based on the belief that our closest living relatives are the African apes. But these two suggestions — one of geography, the other of evolutionary related-ness — are not necessarily connected.

The major premise behind Darwin's pointing to Africa as the seat of human origins and the home of the ancestry of humans and their purportedly closest living relatives, the African apes, is the expectation of finding the remains of ancestors in the same general geographic areas in which the living forms are found today. This premise may at times be supported by the discovery of fossils. But there is no a priori reason why the premise has to be true in every instance. Animals, particularly large animals, migrate — sometimes long, tortuous distances, and sometimes even as an annual event.

In Africa today one finds representatives of *Homo sapiens* as well as two of the large, "manlike" apes, the chimpanzee and the gorilla. And one might even expect to uncover fossils of particular importance to the ancestry of one or another of these animals. It is because of the discovery of early human forerunners in African deposits that we must all concede that the emergence of our more immediate ancestors did, as Darwin predicted, occur in Africa. Although the information is much scantier, we should expect that the modern African apes evolved in Africa as well.

But that is the extent to which Darwinian prediction is justified. Just because humans and African apes are found together in Africa, this does not automatically translate into a statement of relatedness. Our closest living relative could be anywhere now — well, given the constraints of continental drift and major water barriers, almost any-where. And indeed there is another — a third — large, "manlike" ape alive today, living in southeast Asia on the islands of Borneo and Sumatra. This ape is the orang-utan.

I do think that human origins are to be sought in the African past. I also think that our closest living relative is the orang-utan. And, as I will illustrate, there are very good reasons for taking these sugges-tions extremely seriously.

2

In Pursuit of
Human Ancestry

IN 1821, the famous French paleontologist and comparative anatomist Georges Cuvier presented to the scholarly world a new fossil
specimen. He named this ancient animal *Adapis* and incorrectly attributed its affinities to pachyderms. Nevertheless, *Adapis,* which has
since been reinterpreted as being a relative of a subgroup of the living
lemurs of Madagascar, was the first fossil primate to be discovered.

The discovery of *Adapis* and many other fossil mammals was of
no evolutionary consequence to Cuvier. Cuvier, like virtually all natural historians, comparative anatomists, and taxonomists of the early
nineteenth century, was not an evolutionist. Cuvier in particular ad-
hered to a particular school of theological philosophy that approached
the study of earth's layers and the fossils contained in them with an
eye toward demonstrating the biblical interpretation of the history of
the world.

Through his efforts in collecting fossils in and around the limestone
deposits of the Paris basin, Cuvier pieced together a picture of disruption throughout earth's history. The layers of the earth, the strata,
did not flow almost imperceptibly from one to the other. Instead,
there were appreciable discontinuities between the strata as one proceeded up the layers of the basin. These discontinuities were paralleled
in the fossils recovered from each stratum. Rather than a march of
gradually changing organisms throughout the geological column, each
stratum seemed to house a fossil assemblage that was cut off from
its nearest neighbors above and below.

Cuvier and others interpreted the stratigraphic and paleontological discontinuities as evidence of floods or similar catastrophes that occasionally ravaged the earth. In this interpretation, virtually all life would be wiped out during one of these catastrophic events, replaced through acts of creation by the next wave of organisms. The few species that persisted from one stratum to the next, and had thus survived the catastrophe, did so by chance.

Cuvier was on the staff of the Museum of Natural History in Paris. His tenure there overlapped with that of an early evolutionist, Jean Baptiste Lamarck. Lamarck, along with the great French naturalist Georges Louis Leclerc, Comte de Buffon, a major force in the development of the museum and its famous gardens, suggested that humans were related to, or even derived from, an anthropoid primate. The conglomerate Orang Outang was mentioned prominently in this context.

Cuvier and Lamarck were neither friends nor colleagues. And it is ironic that it was Cuvier who gave Lamarck's eulogy, for Cuvier presented the world with an image of a Lamarck who had nothing but foolish ideas. It is common today to read in biology texts how ridiculous Lamarck's ideas were — that offspring can inherit traits acquired during the lifetime of their parents, or that organisms can virtually control their own evolutionary destinies by engendering change through the use or disuse of their organs, for example. What is often overlooked is that these ideas, as well as many others that might sound even more ludicrous to us today, were prevalent during the nineteenth century and were increasingly invoked by Darwin in the course of his many revisions of *On the Origin of Species*.

Sixteen years after Cuvier described *Adapis,* another Frenchman, Edouard Lartet, uncovered a fossil that was thought to have possible affinities with the apes. Lartet discovered the type specimen of this new fossil primate (a "type specimen" is usually the first specimen to receive a new genus or species name) in 1837 near the town of Sansan, France. The specimen consisted of a nearly complete mandible (lower jaw) with all the teeth preserved in it. Since many a type specimen of a new genus or species is based on little more than a fragmentary piece of bone or tooth, it was fortunate that the first fossil ape to be described was based on what we would consider a relatively complete specimen, something to sink your comparative teeth into.

A few years later, the French systematist Henri Marie Ducrotay de Blainville proposed a name for this type specimen: the genus name

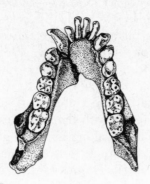

FIGURE 2A   The type mandible of *Pliopithecus.antiquus,* the first fossil thought
to be related to living apes, was discovered by Edouard Lartet in 1837 near the
town of Sansan, France. Some paleontologists think *Pliopithecus* may be related
to modern gibbons.

Pithecus and the species name *antiquus,* which together mean "ancient
ape." Nine years later, in 1849, another French systematist, François
Louis Paul Gervais, changed the genus name to *Pliopithecus.* Today
we refer to the specimen, and to others that are thought to represent
the same species, as *Pliopithecus antiquus.*

The type specimen of *Pliopithecus antiquus* is a mandible about
the size of the mandible of the smallest ape, the gibbon. Modern
gibbons are found throughout southeast Asia and are generally regarded as less closely related to humans than any of the large apes.
The skull of a typical gibbon can easily be cupped in your hand. No
doubt in part because of its similar size, *Pliopithecus antiquus* has
often been thought of as a European gibbonlike fossil. If this association is correct, it would reflect a long period of separation of the
gibbon lineage from the other apes, because specimens of *Pliopithecus
antiquus* come from deposits of the middle of the Miocene epoch,
dating roughly to sixteen million years ago.

The British fossil hunter Arthur Tindell Hopwood made one of his
important discoveries not in the field but in the collections of the
British Museum (Natural History) in London in the late 1920s. There,
amidst a mass of fossil material that had been sent by E. J. Wayland,
director of the Geological Survey in Uganda, to the BM(NH), Hop-

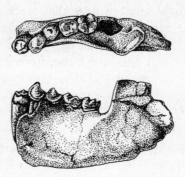

FIGURE 2B Two views of the type mandible of *Dryopithecus fontani*, discovered in 1856 by Edouard Lartet near Saint Gaudens, France. *Dryopithecus* has been an important character in the history of paleoanthropology, being sought, at times, as the ultimate ancestor of both humans and apes.

wood recognized the left upper jaw of what appeared to be an ape. He had to find more.

In 1931 Hopwood set out to East Africa. He discovered nine additional fossil primates, which he announced in a 1933 publication. In that article, Hopwood coined the genus names *Limnopithecus, Proconsul,* and *Xenopithecus* for the new fossil material. *Limnopithecus* was to become recognized as the East African relative of *Pliopithecus.*

In 1856, seven years after the genus name *Pliopithecus* was proposed for the European relative of the gibbons, the first fossil attributable to the large apes was discovered. It was Edouard Lartet who was responsible for this paleontological deed, and it was he who also gave this large fossil ape the genus name *Dryopithecus.* In honor of M. Fontan, who had actually found the specimen near the village of Saint Gaudens, Lartet created the species name *fontani. Dryopithecus fontani* came from deposits that are approximately fourteen million years old.

The type specimen of *Dryopithecus fontani* is composed of portions of the right and left sides of the mandible, from which some of the anterior and posterior teeth are missing. Nonetheless, Lartet, and subsequently others, felt that aspects of the preserved canine teeth, premolars, and molars of the fossil were sufficiently similar in detail to comparable teeth of modern apes that there must be some connection between the extinct and the living species. More recent discoveries of more complete specimens of jaws and teeth, as well as the discovery of a humerus (upper arm bone) from deposits in Germany, have helped create the impression that *Dryopithecus* is a primitive Miocene ape of roughly the same size as the modern chimpanzee.

Darwin must have been aware of the discovery and the implications of the European *Dryopithecus*. However, he made reference to this presumed fossil ape only in 1871, in *The Descent of Man,* twelve years after the publication of the first edition of the *Origin*. Perhaps to avoid bringing upon himself an additional barrage of outrage that a discourse on human evolution would provoke, Darwin made only a passing reference to primate evolution in the *Origin,* and that was in the conclusion of the work: "In the distant future I see open fields for far more important researches . . . Light will be thrown on the origin of man and his history."

As I mentioned in chapter 1, Darwin does tackle the issue of human origins in *The Descent,* first with various arguments that would place the seat of human antiquity in Africa, and second with a conviction that our closest living relatives are the African apes. Darwin does accept the relatedness of "the Dryopithecus" to the large apes and concludes that the human lineage separated from the Old World monkey and ape or "Catarhine" stock quite early, "for that the higher apes had diverged from the lower apes as early as the . . . Miocene period is shown by the existence of Dryopithecus."

Nevertheless, the discovery of *Dryopithecus* in Europe rather than Africa was a nuisance that Darwin had to confront, for a major theme of his work was that "in each great region of the world the living mammals are closely related to the extinct species of the same region." If humans were most closely related to the extant African apes, then, Darwin reasoned, "it is . . . probable that Africa was formerly inhabited by extinct apes allied to the gorilla and chimpanzee." But with only Europe giving forth the fossils whose existence he had predicted and which he needed to connect the living forms of Africa with those of the past, Darwin was forced to admit that "it is useless to speculate on this subject."

While one interpretation of the *Dryopithecus* material was that it represented a Miocene fossil ape, another emphasized the more humanlike qualities of the animal. For example, *Dryopithecus* differs from the modern great apes in that its lower canine tooth is more vertically implanted in its mandible. In the modern apes, this tooth is slanted forward a bit. Even more humanlike than the rather vertical implantation of the lower canine was the fact that in *Dryopithecus* the amount of space in the jaw that the canine and the two premolars behind it occupy is relatively much shorter than it is in living great apes. A reason given for the more crowded dentition of modern humans is that it is correlated with a much shorter jaw and less protruding face than is typical of apes. Perhaps the same general facial and mandibular features characterized the fossil *Dryopithecus*.

The role of *Dryopithecus* in human evolution and its pivotal position in reflecting the antiquity of the human lineage became rapidly implanted in paleoanthropological thought. As a creature of the Miocene, *Dryopithecus* was relatively ancient. It was certainly older than animals from the succeeding Pliocene epoch and it was by no means clear that fossils of other higher primates would be forthcoming from older epochs — the Eocene, for instance.

Dryopithecus became a link, albeit a distant link, between the very distant past and the potential human antecedents of the much more recent past. Without specifically referring to *Dryopithecus*, Thomas Huxley, in his 1863 essay "On Some Fossil Remains of Man," pondered its implications for human evolution: "Where, then, must we look for primaeval Man? Was the oldest *Homo sapiens* pliocene or miocene, or yet more ancient? In still older strata do the fossilized bones of an ape more anthropoid, or a Man more pithecoid, than any yet known await the researches of some unborn paleontologist?"

When the first piece of a Neanderthal was found, in a cave in the Neander valley (hence "Neander"-"thal"), it was thought to represent the remains of a deranged Prussian soldier or Cossack who had wandered off and died in that lonely spot. The thick cranial bone and huge brow ridges of this specimen, and then of other Neanderthal specimens when they were found, were interpreted as deviant derivations — actually, reversions — of once healthy modern humans.

Thomas Huxley was among the first to recognize the evolutionary implications of Neanderthals. Indeed, his essay of 1863 included a long description and discussion of the original Neanderthal skull cap, which, for Huxley, did not represent a modern human fallen from the angels. Rather, "though truly the most pithecoid of known human

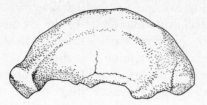

FIGURE 2C   The Neanderthal skull cap, the first Neanderthal specimen found.

skulls, the Neanderthal cranium is by no means so isolated as it appears to be at first, but forms, in reality, the extreme term of a series leading gradually from it to the highest and best developed of crania."

The place of Neanderthal in human evolution became solidified with the discovery in the 1880s of some of the best-preserved specimens ever to be found. These Neanderthal fossils were from the Belgian cave site of Spy (pronounced "Spee") and demonstrated, for the first time, the antiquity of the human species. The Spy Neanderthal skeletons were unquestionably associated with old, Paleolithic stone tools as well as with the bones of extinct animals. Since the Spy Neanderthals looked just like the original Neanderthal specimen in that the cranium was elongate and thick-boned and adorned with massively puffed-up brow ridges, the German Neanderthal skull cap must have been from an equally ancient individual.

The Belgian anatomist J. Fraipont and his geological co-worker M. Lohest described the Spy Neanderthal material in 1886. As Huxley would comment ten years later, "the anatomical characters of the skeletons bear out conclusions which are not flattering to the appearance of the owners." The Spy Neanderthals were short, "but powerfully built." They seemed to have walked somewhat stooped or a bit bent at the knee. Their bones, including their cranial bones, were quite thick and solid, and their mandibles lacked a chin, "that especially characteristic feature of the higher type of man."

While recognizing the human connection of Neanderthal, Huxley and Fraipont and Lohest nonetheless had some difficulty in figuring out how it got to its state of evolution, and especially how long it took. As Darwin had argued, if evolution occurs gradually and by the accumulation of a multitude of minor alterations, then one must postulate an incredible time depth for the human lineage. Thus Huxley was forced to conclude that "if any form of the doctrine of progressive development is correct, we must extend by long epochs the most liberal estimate that has yet been made of the antiquity of Man."

This extrapolation would seem to be inevitable. The logic of the argument and the limited realm of comparison yielded only one reasonable solution for Fraipont and Lohest:

> The distance which separates the man of Spy from the modern anthropoid ape is undoubtedly enormous; between the man of Spy and the *Dryopithecus* it is a little less. But we must be permitted to point out that . . . man . . . has travelled a very great way.
>
> From the data now obtained, it is permissible to believe that we shall be able to pursue the ancestral type of men and the anthropoid apes still further, perhaps as far as the eocene and even beyond.

The antiquity of "man" was extended well into the past through *Dryopithecus*. And it was through *Dryopithecus* that the human lineage was linked to the modern apes. But as Fraipont and Lohest wrote, "between the man of Spy and an existing anthropoid ape there lies an enormous abyss." The evolutionary chain of being was still missing too many links.

The first person to set out to fill in the gaps — to seek and actually find a potential human ancestor — was the Dutch anatomist Eugène Dubois. Like Darwin, Dubois believed that humans must have evolved in one of the two areas of the world in which today we find modern great apes. However, and perhaps because of Dutch control in Malaysia and the islands of Indonesia, Dubois went to southeast Asia rather than to Darwin's place of choice for human origins, Africa.

In 1887 Dubois went to Sumatra and at first worked in a hospital. He spent the following year exploring some of the caves in and around the island but did not find any fossils. In 1889 news reached him of the discovery of a skull from the site of Wadjak in Java, so he went there. Upon his arrival, however, Dubois found that not only was the Wadjak skull not ancient, it was quite recent and representative of more modern Australians.

But Dubois was persistent, and finally, late in 1890, he found the first of four fossil specimens that would prove to be important to his

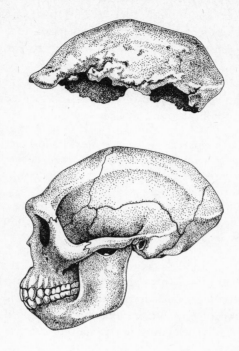

FIGURE 2D   *(Above)* The skull cap found by Eugène Dubois in the 1890s in Java that he referred to *Pithecanthropus erectus.*
   (Below) A reconstruction of how the whole skull of Java Man may have looked. Today, *Pithecanthropus erectus* is recognized as *Homo erectus.*

pursuit of human ancestors. His luck having now changed for the better, Dubois collected in less than two years, and within meters of each other, a calvarium (skull cap), two molar teeth, and a femur. In 1894 he publicly announced his discoveries, and the following year he took his specimens on tour throughout Europe.

All who studied the skull cap agreed that it was certainly not modern human in its features. Nor was this specimen a Neanderthal; Dubois's Javanese fossil calvarium was smaller and less highly vaulted than the skull of a European Neanderthal. "Never yet," Dubois would write, "has there been seen so flat and low a human skull, never yet, outside of the true apes, has so strong a projection of the orbital regions been found." And indeed the supraorbital regions — brow ridges — of Dubois's human antecedent were truly remarkable in their shelflike distention.

The skull cap was the most pithecoid, apelike, of any human fossil so far discovered, and it was also the oldest known human fossil. Without a doubt, this potential ancestor stood in a more obviously intermediate position between modern humans and apes. Dubois was compelled to coin the genus name *Pithecanthropus* ("ape-man") for his Javanese link in the saga of human origins. But for all of its anatomical attributes and its good fortune to come from older deposits than did Neanderthal, *Pithecanthropus,* a predicted "missing link," was to meet with some resistance.

Controversy over *Pithecanthropus* stemmed in large part from Dubois's associating the skull cap with the femur he had found, not just as parts of different individuals of the same species, but as cranial and skeletal parts of the *same* individual. The problem was that although the skull cap was seen as apelike, and thus as fitting right into a developing scheme of human evolution, the femur was wrong, *not* apelike. Rather, the femur was essentially modern in appearance, differing only in being a bit smaller and thicker than the femur of a present-day human. Whoever had had this femur walked as we do, upright and erect.

Undaunted by this apparent inconsistency, Dubois kept the skull cap and femur in association by giving his *Pithecanthropus* the species name *erectus,* in recognition of the existence of an "ape-man" that had walked upright. In 1896 Dubois was still convinced of the reality of his "upright-walking ape-man" and that it "represents a so-called transition form between men and apes, such as paleontology has often taught us to recognize between families of mammals." *Pithecanthropus erectus* was, for Dubois, "the immediate progenitor of the human race." But what is most curious about Dubois is that he found his comparisons of *Pithecanthropus* to be most favorable with the gibbon, and not one of the great apes, so that in later years he came to suggest that his early "ape-man" was a giant gibbon instead. Despite his own uncertainties about its affiliation, *Pithecanthropus erectus,* "Java Man," was eventually accepted, femur and all, and came to be known as *Homo erectus.*

By the turn of the twentieth century, almost all doubt about the existence of a fossil record documenting the evolution of our own lineage had disappeared. Neanderthals were becoming quite well known from sites uncovered throughout Europe. One of the most famous Neanderthal sites, a grotto in France near the town of La Chapelle-aux-Saints, yielded in 1908 the remains of an older, apparently male individual. Remarkably, this individual had survived beyond his early years, even though his bones showed a history of various illnesses, including severe periodontic problems that had resulted in the loss of all but a few of his teeth. For the first time, modern humans got a glimpse of community and cooperation — which must have been the reason for the prolonged survival of this Neanderthal — in an extinct member of their own evolutionary past.

But the question was, and still is, how closely related was Neanderthal to modern humans? The French physical anthropologist Marcelin Boule thought that "the skeletons . . . do not justify a generic distinction." But Boule would distinguish Neanderthal at the species level from modern *Homo sapiens* and place it "exactly between the Pithecanthropus of Java and the more primitive living races." This act, however, would not "imply . . . the existence of direct genetic descent." While it is important historically that *Pithecanthropus* and Neanderthal came to be recognized as having a place in our evolutionary past, there still remained a lot of missing links between the human and ape lineages.

Darwin had suggested that there should have been a graded series of forms transitional between the knuckle-walking apes and the fully erect, bipedal *Homo sapiens*. He also sought to elucidate the conditions necessary for the "acquirement of higher mental qualities," those "mental powers" so characteristic of modern humans, especially the more "civilized" groups. Thus were delineated the two major features that came to be recognized as unassailably human: habitual, bipedal locomotion, and the possession of a large and complex brain. The question was, however, which feature evolved first. Would the fossil record show us earlier evidence for the development of bipedalism? Or would the first advance away from the "pithecoid" (apelike) state involve the expansion of the brain, the anatomy of "imagination, reason, abstraction, language"? Along with the expansion of the brain would be seen a "humanization" of the skull, with it becoming larger, rounded, and generally more lightly built than in an ape.

Dubois's Javanese fossil, *Pithecanthropus erectus,* represented one solution to the problem. If the skull cap, with its low contour and heavy brow ridges, and the femur, with its modern human morphology, did go together, then the picture of human evolution involved first the development of bipedal locomotion and then the expansion of the brain and the remodeling of the cranium. Judging by Java Man, our ancestors became human from the neck down while remaining apelike in their heads. Expansion of the brain and the associated cranial enlargement and lightening up came later. A common scenario given to explain this series of events was that bipedalism freed the hands for functions more "thought-provoking" than merely getting about, such as making tools, and thus became a selective agent that influenced the evolution of the brain. However, not too long after the discovery of *Pithecanthropus,* evidence of the alternative solution, of the evolution of brain before body, emerged — in England, no less, one of the least likely seats of early human evolution, at a site called Piltdown.

The first piece of the Piltdown puzzle was discovered in 1911 by Charles Dawson. With the aid and support of Sir Arthur Smith Woodward of the BM(NH), Dawson eventually recovered from the Piltdown gravels nine cranial fragments and a large portion of a right mandible in which only the first and second molars were preserved. The cranial fragments could be reconstructed to form much of the

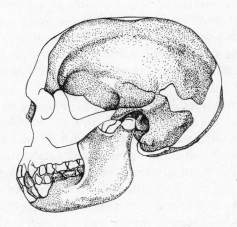

FIGURE 2E A composite reconstruction of the skull and jaw of Piltdown Man, most of which was discovered in 1911 by Charles Dawson, who subsequently found the canine in 1914. (The image of the canine and mandible are reversed for the drawing.)

back portion of the skull; nothing was known of the facial region
except a small portion of the upper rim of the left orbit (eye socket).
What struck the experts who studied the specimen was that the skull
was lightly built and highly vaulted and that brow ridges were absent.
This, of course, was in marked contrast to the thicker-boned, flatter
cranium of Neanderthal and *Pithecanthropus,* whose orbital regions
were topped by massive brow ridges. The skull — or what remained
of it — looked essentially modern.

In contrast to the skull, the mandible, even though not complete,
was obviously not as human-looking as it was reminiscent of ape
morphology. Most apelike were the shape of the preserved midline
region of the jaw and the alignment of the premolar and molar regions.
However, Smith Woodward was quite struck by the nature of the
two molar teeth, which he felt were "distinctly human" in their overall
appearance, but "of the most primitive type," being apelike in their
relative narrowness.

The most curious aspect of the lower molars of "Piltdown Man,"
as this newly crowned human ancestor soon came to be called, was
that although their chewing surfaces were quite worn, very little of
the underlying material, the dentine, was exposed. Only in animals
that develop a very thick layer of enamel over the softer dentine
would this be possible. Among primates, the possession of thick
molar enamel is quite uncommon: only one species of New World
monkey and only one species of Old World monkey develop thick
enamel on their molar teeth. Otherwise, among extant primates,
relatively thick molar enamel is present only in humans and orang-
utans.

The Piltdown mandible had the same coloring as the skull; for
them to become so equally mineralized, they must have been in the
ground for the same amount of time. Although the head of the mandible (the part that comes into contact and articulates with the base
of the skull just in front of the ear region) was broken off, it nevertheless seemed that the mandible could have meshed with the skull.
The association of the skull and mandible, perhaps as remnants of
the same individual, seemed secure and was accepted by many, especially British, anatomists and paleontologists. However, one prominent critic of this association of skull and mandible, the German
anatomist and paleoanthropologist Franz Weidenreich, thought that
the jaw was not from anything human but from an orang-utan.

But the euphoria of England at being able to claim *the* human ancestor as one of its proud sons was overwhelming. Smith Woodward created a new genus and species, *Eoanthropus dawsoni,* to accommodate the Piltdown skull and mandible. The species name honored the discoverer, Charles Dawson. *Eoanthropus* means "dawnman" and was coined because Smith Woodward was convinced that modern *Homo sapiens* had evolved directly from the Piltdown type of early human. For Smith Woodward, the more brutish-looking Neanderthal became even more the "degenerate offshoot."

The following year, 1914, an isolated canine tooth was found in the Piltdown gravels, virtually in the same spot that had yielded the skull and mandible. It was immediately and without question interpreted by Smith Woodward as having come from the earlier-found mandible. The canine tooth was quite worn down, as the molars had been. Unlike the molars, however, the wear on the canine was pervasive, exposing quite a bit of dentine below. Smith Woodward concluded that the canine "must have been well used for a considerable period" and thus must be from something humanlike, because its extensive wear could only mean that it had erupted into the jaw "before the second and third molars, as in Man — not after one or both of these teeth, as in the Apes."

A place, if not *the* place, of Piltdown Man in human evolution seemed assured, even though the material generated quite a bit of heated debate and discussion. The unexpected and almost incongruous association of a big-brained, humanlike skull with suggestively human teeth implanted in a much more apelike jaw intrigued and puzzled students of human evolution for over four decades. But with the application in the 1950s of new analytical chemical techniques for determining the antiquity of bone and the use of a high-powered microscope to identify artificially induced scratch marks on the teeth, the paleontological pride of England, Piltdown Man, was shown to be a fraud. The whole thing had been a forgery, a carefully and masterfully carried-out plan to create a missing link, one that could be viewed as an acceptable alternative to Java Man.

The Piltdown skull and jaw were not ancient; they had just been stained chemically to look old. The skull was from a modern and relatively recently deceased human. The jaw, that of an orang-utan, had been broken in just the right places to make disproving its association with the skull difficult. The molar teeth had been filed flat, to accentuate their humanlike appearance. The isolated lower canine had also been modified, to give the impression of its erupting in the

jaw early in life, as in humans, rather than as the last of the adult
teeth, as is the case with apes.

When I was a graduate student, I spent a year studying at the
BM(NH) under the wing of Theya Molleson, who had worked closely
with Kenneth Oakley, the person who had developed and run the
important tests that eventually revealed the composite nature of Piltdown Man. When I was there, in 1971, Kenneth Oakley was quite
old and came infrequently to the museum. I met him only once. He
has since died.

When I returned a few years ago on one of my research trips to
the BM(NH), Theya told me that Oakley had promised to leave her
a document exposing the perpetrator, or perpetrators, of the fraud —
to be disclosed only after his death. However, the document never
turned up. The secret died with Kenneth Oakley; we were left only
with a history of some scholars accusing Dawson and others taking
on the distinguished Sir Arthur Smith Woodward. Most recently, even
the seemingly unassailable French cleric and paleontologist Teilhard
de Chardin, who happened to find the canine during a visit he made
to the excavations of the Piltdown site, has been the target of accu-
sations of complicity in this paleontological lie. The situation is certainly frustrating, especially since Theya has told me that she had
some biochemical analyses done which not only supported the identification of the isolated canine as a tooth from an orang-utan but
also indicated that the canine had originally come from the Piltdown
mandible. Whoever planted the skull and mandible in the first place
appears to have anticipated having to produce some additional "evidence" at a later date.

We will probably never have much more concrete information
about the Piltdown forgery. But although Piltdown Man no longer
holds for us any insight into the more recent events of human evolution, it does quite unexpectedly provide us with a tantalizing clue
about our more distant relations. The Piltdown forgery could not
have been perpetrated if the teeth of any extant primate other than
an orang-utan had been used. The mandible of a large chimpanzee
or a smallish gorilla, depending on how it was broken, would probably
have accomplished the same job as the orang-utan jaw in terms of
giving an overall apelike impression. However, the teeth of either of
the African apes could not have been fiddled with to appear humanlike. The molar cusps of a chimpanzee and especially of a gorilla are

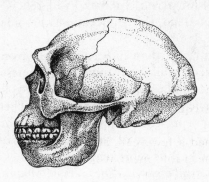

FIGURE 2F Franz Weidenreich's reconstruction of Peking Man, originally called *Sinanthropus pekinensis,* from the site of Chou Kou Tien, China. Today, *Sinanthropus pekinensis* is referred to the genus and species *Homo erectus.*

higher than ours or those of an orang-utan, and the molar enamel of an African ape is thin. Any filing of their teeth would have exposed quite a bit of dentine, as is the case in the normal wear and tear on a chimpanzee's or a gorilla's teeth. But an orang-utan's molars are sufficiently humanlike to be touched up a bit and fool an array of experts.

Either in spite of, or because of, the claims that were being made about *Eoanthropus dawsoni,* pursuit of human ancestors continued actively through the 1920s. Driven by the same motivations as those that had earlier sent Dubois to Sumatra and then Java, Davidson Black, a British anatomist, had gone to China with the express desire of finding human ancestors. In 1927, two years after his appointment as professor of anatomy in the recently established medical school, he did. The cave site of Chou Kou Tien, near Beijing (Peking), yielded a very humanlike tooth. Black made this tooth the type specimen of his new early human, which he called *Sinanthropus pekinensis,* "China man from the region of Pekin."

Fortunately, "China man from the region of Pekin," or just "Peking Man," did not remain known from only one tooth. During a ten-year period of excavation — until the war between Japan and China caused work to cease — various skulls and mandibles and numerous isolated teeth were uncovered. In addition to these fine fossils came the discovery of associated primitive chopping tools and the earliest archeological record of the purposeful making of fire. With a date of about half a million years ago, this still remains the oldest unquestionable record of manmade fire.

Although a bit larger than *Pithecanthropus* from Java, *Sinanthropus* was recognized by all who studied the Chinese fossil material as very similar to the smaller form: the cranium was low in contour and bore rather prominently distended brow ridges. There was no question that *Sinanthropus* represented a primitive type of human. Debate was usually restricted to the issue of whether *Sinanthropus* and *Pithecanthropus* should be kept as separate genera or whether they should be collapsed into the same genus. These days, *Sinanthropus* and *Pithecanthropus* are considered to represent geographical and temporal variants of the same species. Except for historical reference, these antiquated genus names are no longer used. They are not considered taxonomically valid. We now recognize *Pithecanthropus* and *Sinanthropus* as belonging not only to the same genus but also to the same species, *Homo erectus*.

One of the greatest disasters in the history of human paleontology was the loss of the Chou Kou Tien specimens during the Sino-Japanese War. The specimens may have been blown up as they were being transported by train, or they may have sunk with the ship on which they were put for safe passage. Fortunately, Franz Weidenreich, who took up the Chou Kou Tien project after Davidson Black died in 1934, had made extensive notes as well as molds of the fossil material.

In 1924 in South Africa, a young anatomist, Raymond Dart, was presented with a small fossilized skull that had been blasted out of the limestone deposits of the site of Taung. Skulls of fossil monkeys — baboons, actually — were already known from these deposits. But this skull was different. Dart recognized immediately that this new find was, as he called it, an anthropoid, not a monkey. He published notice of the discovery and his interpretation in the February 1925 issue of the prestigious British journal *Nature*. The world of human evolutionary studies has never been the same since.

The specimen consisted of much of the front of the skull and mandible of a child who, by modern growth landmarks, was probably not older than six years of age at death. In fact, a recent study has concluded that the Taung child could have been as young as three years of age at death. The milk teeth (the first set) were present, and

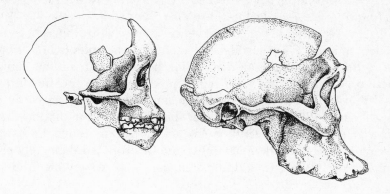

FIGURE 2G   *(Left)* The partial skull of the child from the South African fossil hominid site of Taung. Raymond Dart, who in 1925 presented this specimen to the world in the journal *Nature,* named this hominid *Australopithecus africanus.*

(Right) The adult skull that Robert Broom discovered at the South African fossil hominid site of Sterkfontein. First placed in the genus *Australopithecus,* with its own species, *transvaalensis,* this specimen was later given its own genus name and thus became known as *Plesianthropus transvaalensis.* Eventually this skull was recognized as an adult *Australopithecus africanus.*

the first permanent molar had also come into place. Aside from the well-preserved facial region, the rest of the skull was never recovered, but a natural cast of a large portion of the child's brain was found. This "endocast" was produced by dissolved limestone dripping into the brain case and forming a small deposit inside the bony chamber. Thus while the bone was becoming mineralized, a partial cast of the inside of the skull was also being made.

It was the analysis of this partial endocast that first prompted Dart to realize "that one was handling in this instance an anthropoid and not a cercopithecid ape." (At the time Dart was writing, the Old World monkeys, or cercopithecids, were referred to as apes, but they were known to be different from the real — or, as they were called, anthropoid — apes.) Upon considering all that had been found of this individual, Dart was forced to conclude that the whole cranium, the dentition, and the mandible were "*humanoid* rather than anthropoid." As was typical of this rather bold young scholar, Dart even went so far in this premier publication in *Nature* as to claim that "the specimen is of importance because it exhibits an extinct race of apes *intermediate between living anthropoids and man.*"

Dart considered his specimen "a creature well advanced beyond modern anthropoids in just those characters, facial and cerebral, which are to be anticipated in an extinct link between man and his simian ancestor." Thus, Dart's fossil represented a manlike ape, in contrast to Dubois's *Pithecanthropus,* which had been characterized as an "apelike man." The intermediate evolutionary position that his fossil from Taung held between humans and apes Dart thought was deserving of a new taxonomic family, and he proposed the barely pronounceable family name Homo-simiadae. This taxonomic addition was never accepted.

Dart also proposed "that the first known species of the group be designated *Australopithecus africanus,* in commemoration, first of the extreme southern and unexpected horizon of its discovery, and secondly, of the continent in which so many new and important discoveries connected with the early history of man have recently been made, thus vindicating the Darwinian claim that Africa would prove to be the cradle of mankind." Despite the efforts of many prominent physical anthropologists and paleoanthropologists to suppress the whole matter and discredit Dart, *Australopithecus africanus* remains as a testimony to the insights and courage of this scientist.

One of the more profound effects of Dart's recognition of this very primitive humanlike form was that it, as Dart was proud to proclaim, vindicated Darwin. With all the discoveries of this or another region's fossil "man," Africa was then seen as anything but the cradle of humankind. The Taung child changed all of that. It was the most apelike humanlike specimen to ever be uncovered. In fact, it was so apelike that many eminent scholars claimed the Taung skull was really some sort of anthropoid in general, or even a new species of fossil chimpanzee or gorilla. But Dart held his ground and retained his conviction that he had indeed found a true link between humans and apes.

Dart's argument was essentially a rephrasing of Darwin's earlier opinion that lush, tropical countries, such as those that were yielding the Asian fossils, could not provide the evolutionary impetus necessary for the development of a human sort of creature. Life there was just too comfortable. What was needed were the dangers of such a place as Africa. As Dart phrased it:

In the luxuriant forest of the tropical belts, Nature was supplying with profligate and lavish hand an easy and sluggish solution. For the production of man a different apprenticeship was needed to sharpen the wits and quicken the higher manifestations of intellect — a more open veldt country where competition was keener between swiftness and stealth, and where adroitness of thinking and movement played a preponderating role in the preservation of the species . . . Southern Africa, by providing a vast open country with occasional wooded belts and a relative scarcity of water, together with a fierce and bitter mammalian competition, furnished a laboratory such as was essential to this penultimate phase of human evolution.

Dart certainly stirred things up in 1925. But it is quite possible that his ideas would have stayed squelched forever had he not received the support of the eminent vertebrate paleontologist Robert Broom, who was convinced by Dart's argument that *Australopithecus africanus* was indeed humanlike. Just as Darwin had his bulldog in Huxley, so Dart had his in Broom.

Although Robert Broom defended Dart and his interpretation of the Taung child, he did not become active in human paleontology and the pursuit of human fossils until 1934. Up until that time, and even though he had become quite well known for his work on fossil reptiles, Broom was occupied principally with being a physician. But in 1934 he accepted a position in the Transvaal Museum, Pretoria, South Africa, and immediately immersed himself full-time into matters paleontological.

In 1936 two of Dart's students told Broom of a cave about thirty miles from Johannesburg which was quite rich in fossil mammals. Unfortunately, the limestone deposits of this cave, Sterkfontein, were being quarried for fuel for lime kilns. The richness of the limestone in fossil material could not have been entirely unknown, for Broom cites a travel guidebook in which was written, "Come to Sterkfontein and find the missing link." And so, to paraphrase his own account of what happened, Broom went, and he did.

In August 1936, Broom and his co-workers found a large portion of the skull of an adult specimen. Because of its overall similarity in form to the Taung child, this adult specimen was placed at first in the same genus, *Australopithecus*. However, the adult specimen from Sterkfontein was given its own species, *transvaalensis*, to denote its having been discovered in the South African Transvaal. But because of differences that were later seen to exist between the Sterkfontein

and Taung specimens, Broom created a new genus for the adult specimen, which then became known as *Plesianthropus transvaalensis*.
Although a lot of fossil material was to emerge from the deposits at
Sterkfontein, the *Plesianthropus* skull remained the best and most
complete specimen until 1947, when an even better-preserved skull
was discovered.

In June 1938, Broom was presented with a portion of the upper
jaw of an "ape-man" much larger than the obviously more lightly
built and gracile type found at Sterkfontein. A schoolboy who gave
tours of the area's caves had found the new specimen at the site of
Kromdraai. Upon being taken to the site, Broom had the dirt from
which the specimen came sieved, and he eventually recovered bits and
pieces that constituted a large portion of a skull and a mandible.
These fossils represented a species that was not only much larger but
also much more robust and powerful-looking than the "ape-man"
from Sterkfontein. This more robust form had a much flatter face
than the rather duck-billed Sterkfontein specimen. The teeth of the
Kromdraai "ape-man" were huge and their grinding surfaces tablelike.

Broom created a new genus and species to accommodate this robust
type of "ape-man," *Paranthropus robustus*. His habit of creating new

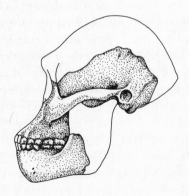

FIGURE 2H   Robert Broom's reconstruction of the robust "ape-man" from
the South African site of Kromdraai. This type of hominid Broom called *Paranthropus robustus*, but most paleoanthropologists consider it a species of *Australopithecus, Australopithecus robustus*. Subsequent discoveries of South African
robust *Australopithecus* demonstrate that this hominid looks more like the very
robust form illustrated in figure 21 than the image in Broom's reconstruction.

genus names every time he found a new fossil met with considerable resistance then and for the next few decades. But one thing is now clear: we cannot deny the coexistence a few million years ago of two types of early human predecessors, a gracile form and a robust one. If they do not represent two different genera, these two distinctly different early hominids, or human-related forms, are certainly representatives of two different species.

Why was the idea so troubling that there existed at the same time, and even in the same place, two different hominids? The problem was that all sorts of notions — not all or perhaps even most of them tangible — had become tied up in the definition of what it took to become human or even to be admitted broadly to our own taxonomic family, Hominidae. Our "true" antecedents, the early hominids, were supposed to be primarily carnivorous, meat-eating, in their diet. They were supposed to be bipedal and reasonably good at it. And they were supposed to have relatively large brains and the mental powers necessary to use and even manufacture tools with which they could manipulate their environment, as we do. Such a combination of characteristics would have to have been rare.

But if these features were so unusual, how could there exist more than one hominid at any one time with these attributes? Some physical anthropologists tried to argue that the robust early hominid was in fact the male, and the more lightly built and gracile early hominid the female, of the same species of *Australopithecus*. But then there would have been only males at Kromdraai and only females at Sterkfontein. With the later discovery of the robust type of early hominid at the nearby site of Swartkrans and of the gracile type at a more distant place, Makapansgat, the idea of there being sex-specific sites became unwieldy, and scientists had to accept the notion that there were two types of early hominid.

The acceptance of the contemporaneity of the two types was not as readily forthcoming, however. There were attempts to prove that the gracile type preceded, and thus could have given rise to, the robust form, and there was an occasional claim that in fact it was the other way around. Since the South African cave sites are of limestone, and not the volcanic débris typical of the East African early hominid sites, the rock as well as the fossilized bones contained in it cannot be analyzed chemically to determine age. Therefore, other, more fanciful criteria were usually brought into arguments about who of these early hominids preceded whom. It was not possible to make definitive rebuttals until chemical dating techniques were applied to the hominid-bearing volcanic deposits of East Africa. Louis and Mary Leakey at Olduvai Gorge, Tanzania, F. Clark Howell and C. Arambourg at Omo, Ethiopia, and Richard Leakey at East Turkana, Kenya, amassed collections of *Australopithecus* that demonstrated, once and for all, the coexistence of robust and gracile forms.

Even when the contemporaneity of two types of early hominid could not be disputed, some still felt uneasy with it. Scenarios explaining how this could be "resolved" the issue by diminishing the qualities of one of these early hominids, leaving the one that seemed more like us in our line of descent. Thus, for example, the robust form was supposed to be an inefficient bipedalist at best. In fact, it was even suggested that the robust *Australopithecus* may have waddled. It could never have made tools and probably did not even know to use objects for such purposes. And this robust species of early hominid was supposed to subsist primarily on roots, seeds, tubers, and grasses, not meat.

But the study of the fossilized fragments of our predecessors has demonstrated that robust *Australopithecus* was just as adept at walking on its two legs as was its contemporary, the gracile *Australopithecus*. Apparent differences between the two hominids in the shape of the pelvis, which had earlier been interpreted as reflecting different bipedal abilities, were really due to the different sizes of the two. A large *Australopithecus*, or a large anything, would not have all of its body parts in the same proportions as a smaller version of the same type of organism. Thus, many differences between the two species of *Australopithecus* in the shapes and relative proportions of their teeth could also be attributed to differences in size. Robust *Australopithecus* had larger, more tablelike grinding teeth because it was larger than the gracile form, not necessarily because its diet was different. The new interpretation was that both varieties of *Australopithecus* were unconstrained dietarily: they ate a bit of everything when it was available. Our early relatives were omnivorous.

A few stalwart holdouts aside, the general consensus at present is that there are two species of *Australopithecus* represented in the South African cave deposits and that these sites range in age from over three to less than one million years old. All of the gracile hominid material is included in Dart's genus and species *Australopithecus africanus*, and the Taung child skull is the type specimen. In addition to Taung, the other typically gracile *Australopithecus* sites are Sterkfontein and Makapansgat. The other species is *Australopithecus robustus*. Its type specimen came from Kromdraai; it and Swartkrans are known as the robust sites.

A popular scenario over the years has been that the genus *Homo* evolved specifically from *Australopithecus africanus*. In presenting the scenario, some physical anthropologists would retain the genus *Paranthropus* for the robust species *robustus* and place the gracile species *africanus* in our own genus, *Homo,* thus doing away altogether with the genus *Australopithecus*. This taxonomic move is based on what would be perceived as the greater evolutionary distance of the robust form from the main line of human descent.

Such an argument is not uncommon in paleontological and systematic circles. But it is not an argument based on phylogeny, the evolutionary relationships of the organisms concerned; rather, it is an argument whose concern is with taxonomy, the naming of the organisms. The number of species one includes in the genus *Australopithecus* is not crucial to resolving the problem of who is related to whom. But the confusion of taxonomy with phylogeny has pervaded the study of human evolution and will probably continue to do so. Indeed, a history of human evolutionary and paleontological studies is really a history of the creation of genus and species names, bestowed by a series of discoverers who thought they had found *the* ancestor or missing link.

The tendency toward giving every new specimen a new and completely different taxonomic identity was characteristic of the late Louis S. B. Leakey. In his judgment, almost everything that came out of the deposits of his site, Olduvai Gorge, in East Africa, was unique. Such an attitude had the unfortunate effect of generating a huge amount of debate on whether each particular specimen actually deserved the special status that a different taxonomic handle immediately carried with it, diverting attention from the important task of studying the specimens themselves.

In 1959, at Olduvai Gorge, Louis and Mary Leakey discovered an almost complete skull of a huge *Australopithecus*-like hominid. It was definitely a robust type, but larger and much more robust, with much more massive cheek teeth (molars and premolars), than any *robustus* from any known South African site. Louis Leakey felt that this specimen was sufficiently distinct from the South African *robustus* to

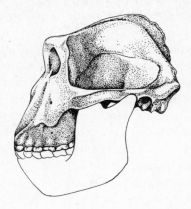

FIGURE 21   The cranium of the hyper-robust hominid from the East African
site Olduvai Gorge; the jaw is drawn from a specimen from a different site.
Discovered in 1959 by Louis and Mary Leakey, this hominid was given the new
genus and species names *Zinjanthropus boisei*. Since they are in many ways larger
versions of the South African robust *Australopithecus,* these hyper-robust hominids were referred to *Australopithecus boisei.*

warrant a new genus and species. This new hominid, *Zinjanthropus
boisei,* which was also known as "Nutcracker Boy" because of its
massive grinding teeth, was truly a super- or hyper-robust type of
early hominid.

F. Clark Howell at Omo, in Ethiopia, and Richard Leakey at East
Turkana, in Kenya, just south of Omo, provided additional specimens
of this hyper-robust early hominid. With increasingly good samples
of the East African hyper-robust hominid and continued comparisons
to the South African *Australopithecus robustus,* it became apparent
that these two forms were evolutionarily very close to each other —
so close, in fact, that they should be considered phylogenetic (evolutionary) sisters. The genus name *Zinjanthropus* was dropped and
the East African hyper-robust hominid included in the genus *Australopithecus* as *A. boisei.* Informally, however, this particular specimen
is still referred to as "Zinjanthropus" or just "Zinj."

In 1960 Louis Leakey discovered most of the bones of a foot of a
hominid. What was striking about the find was that when the met-

atarsals (the bones above the toes) were reassembled in the correct anatomical positions, the foot appeared more humanlike than anything else. For example, the big toe was not oriented outward as a "thumb," which is what it looks like in apes. Rather, this fossil foot had a big toe that was more aligned with the digits of the other toes, as in our own foot. There was also indication from other bones of the foot that there had been an arch, which is for us a shock absorber as we pound around on two legs. The individual who had this foot must have walked bipedally, as we do. But even though these foot bones had been found in the same geological bed as the Nutcracker Boy, Leakey and colleagues could not believe that such a hyper-robust hominid could have had such a modern, human-looking foot.

Later in 1960 Leakey found an almost complete mandible and large pieces of right and left cranial bones. The mandibular fragment was obviously from a juvenile individual, and given the state of development of the right and left cranial bones, all elements were thought to have come from the same individual. By modern aging criteria, especially the state of eruption of the teeth, the mandible would have been from an individual not older than twelve years. The teeth themselves were smaller than the massive teeth of the essentially contemporaneous Zinjanthropus. The cranial bones showed the unfused edges characteristic of a juvenile individual. They were also thin and devoid of the thickened scars of heavy muscle attachment, and were thus distinguished from the massiveness of the bone of the skull of Zinj. The thin cranial bones would also later yield a reconstructed cranial capacity of 600 cubic centimeters, which is appreciably larger than the 530-cubic-centimeter volume calculated for the brain case of Zinj.

The juvenile mandible and cranial bones had been found in the same stratigraphic level as the foot bones and the Zinj skull. And since the thinner-boned, smaller-toothed mandible and the gracile cranial bones were of an apparently more humanlike individual than Zinj, Leakey and his colleagues received the foot bones as a component of their long-lost skeleton. Thus, the mandible with its teeth, the cranial bones, and the foot bones were thought of as representing a markedly different contemporary of the hyper-robust Zinjanthropus. The presence at Olduvai Gorge of a more humanlike form seemed incontrovertible when stone tools, pebbles that had obviously been modified in shape, were found in the same level as the fossils. No one

considered even for a moment the possibility that the brutish Zinj
had made these stone tools. It seemed obvious that if you couldn't
stand upright and walk bipedally, you certainly couldn't manufacture
tools.

In 1964, in a publication in *Nature*, Louis Leakey, Phillip V. Tobias
(of the University of Witwatersrand in Johannesburg), and John R.
Napier (of the Royal Free Hospital of Medicine in London) rewrote
the genus *Homo* to accommodate this non-Zinj material. The new
species, called *Homo habilis* ("Handy Man") to emphasize the ability
of this hominid to make, not just use, tools, became Leakey's direct
ancestor of the other species of the genus *Homo*.

As I mentioned earlier, one of the unassailable bastions of "humanness" was the development of a large brain. The famous British
anatomist Sir Arthur Keith had portrayed the gap between humans
and the rest of the primates as the "Cerebral Rubicon," across which
any potential ancestor must pass before it could be considered a
member of the human lineage. Depending on which scholar you listened to, the minimum brain size needed for its possessor's inclusion
in the genus *Homo* was 700, 750, or 800 cubic centimeters. However,
the reconstructed cranial capacity of the new Olduvai material, the
direct human ancestor, was only about 600 cubic centimeters. In order
for Handy Man to be included in the genus *Homo,* the Cerebral
Rubicon had to be lowered at least as far as 600 cubic centimeters.
It was, and thus was created the first member of the genus *Homo,*
that genius of a genus, replete not only with bipedalism but with full
mental powers.

The Leakeys' tradition of spectacular discoveries and idiosyncrasies
was continued by their son, Richard, in an area north of Olduvai
Gorge. He found a promising area for fossil hunting while flying his
own plane on a survey of the badlands of Kenya in and around the
east shore of what is now called Lake Turkana. Throughout much
of the 1960s and into the 1970s, Richard Leakey and his crew found
an amazing number of relatively complete skulls, jaws, limb bones,
and of course isolated teeth. Two specimens of hyper-robust *Australopithecus* provided for the first time indisputable evidence of sexual dimorphism in this species of early hominid. Both of the fairly
complete skulls display all the diagnostic features of the hyper-robust
form, but one of the skulls is almost twice as large as the other. The
larger skull has massive cheek bones and even a large crest along the
midline of the skull, which would have provided additional area for

the attachment of very large chewing muscles. These differences, in overall size and in the development of morphological features, are precisely the differences one sees between male and female gorillas, for example. Sexual dimorphism in hyper-robust *Australopithecus* was much more marked than in its contemporaneous cousins.

Another type of hominid was also found at East Turkana. This large-brained, non-robust-looking skull came from deposits that were supposed to be older than 2.65 million years of age. Although Richard Leakey and his colleagues preferred to refer to this specimen by its catalogue number (KNM ER 1470), they were not reluctant to include it in the genus *Homo* because its cranial capacity, estimated to be about 700 cubic centimeters, was quite large for an early hominid. With its apparent age, this skull was supposed to represent the oldest known record of the genus *Homo* and rivaled *Australopithecus* in its antiquity. Richard Leakey followed his father in deriving subsequent hominids from his ancient *Homo* and in relegating the *Australopithecus* group to a side line of human evolution.

But this was not the end of it. Almost immediately the age of the East Turkana deposits was questioned. Additionally, the nonhominid animal remains from these deposits were not the same as those of deposits of presumed equal age from F. Clark Howell's hominid site further north, Omo. The pigs, for example, did not match up. Instead, the fossil pigs from East Turkana matched Omo fossil pigs from much younger deposits.

The dust settled; Richard Leakey's hominids were found to be not older than two million years of age. In fact, their antiquity is comparable to the age of the deposits of Olduvai Gorge that yielded similar hominids. In no way did this finding diminish the scientific value of Richard Leakey's discoveries, but it reopened the door for any and all other candidates that might be proposed for the position of our earliest ancestor.

Almost like flies to honey, paleoanthropologists were drawn to deposits much further north, in Ethiopia, in an arid, barren, and hostile region known as the Afar Triangle. This region, surveyed first by a Belgian and an American geologist, is an area where three continental plates are in contact and thus might provide information about the geological history of that part of Africa.

In 1972 or so, the American geologist John Kalb came to the American Museum of Natural History in New York City to give an informal presentation on the geology and paleontology of the Afar

Triangle, and in particular of a special area, Hadar. I happened to be at the museum that day and was one of the lucky few to be audience to a fascinating lecture and slide presentation. The geology was amazing. The fossils, however, were astonishing. Virtually entire skeletons of fossil baboons, elephants, pigs, and almost any other extinct creature of that time period were preserved and were exposed, because of erosion, on the surface of the deposits. The prospects for major finds in all areas were phenomenal.

In the spring of 1974, Clifford Jolly, a physical anthropologist at New York University, organized an international conference on hominid evolution. Just about everybody who had anything to do with human paleontology was there. Somewhere along in the program, Jolly introduced the next speaker, Donald C. Johanson. Most of us had never heard of this young man, but that made no difference. After captivating the audience with a barrage of slides on the geology of the Hadar region and the way in which skeletons of fossil animals were lying on the surface just waiting for someone to collect them, Johanson flashed on the screen the bones of a right knee joint, not of a pig or baboon but of a hominid.

At the time it seemed somewhat excessive to have spent so much time building up to a knee joint — just the lower part of the femur and the upper part of a tibia — even though it was the knee of a bipedal individual. For most of the conference, we had been treated to displays of skulls, mandibles, and plenty of skeletal bones. But Johanson would go back to Hadar and return first with an almost complete skeleton of an early hominid, which he named Lucy, and then with the remains of at least thirteen fossil hominids, ranging from child to adult, which he called the First Family. What Johanson also came away with was an age of about three and a half million years for Lucy. Now *he* carried the baton of human ancestry, and Lucy was it.

Study of the skeleton of Lucy quickly revealed that she had been bipedal. Johanson could thus claim having the earliest record of bipedalism. For some of us, this was not terribly earth-shattering because all other *Australopithecus* (Lucy and other specimens were to be recognized as a species of *Australopithecus, A. afarensis*) were found to have been bipedal. That is, one would have predicted that Lucy would have been bipedal.

What *was* surprising, however, was the discovery that Lucy was much more primitive in her teeth and skull than any other bipedal early hominid. Her skull was tiny, and her jaws and especially her teeth were large. To be a bit general about it, Lucy was humanlike in her postcranial skeleton but apelike above the neck. Given her antiquity, Lucy was a perfect ancestor. She was in the right place in time and seemed primitive enough, and thus presumably apelike enough, to have given rise not only to the other species of *Australopithecus* but to the genus *Homo*.

To be sure, the inevitable debates have ensued. The deposit that yielded Lucy may not be as old as originally thought. Perhaps not all the specimens assigned to *Australopithecus afarensis* belong there, and maybe the species *afarensis* is invalid or even unnecessary. But the discovery of Lucy has served many important purposes nevertheless. One of them is the inadvertent validation of molecular approaches to the issue of human and ape relatedness and the question of when apes diverged from humans, or vice versa.

Since the 1960s, Vincent Sarich and Allan Wilson of Berkeley have been analyzing the immunological reactions of various molecules among primates as a way of determining the genetic, and thus the presumed evolutionary, distances of one species from another. These molecular systematists have also tried to calibrate what they call a "molecular" or "immunological" clock to determine the age of these branching points.

For over a decade, Sarich and Wilson were at loggerheads with paleontologists. The paleontologists claimed that the human and ape lineages diverged over eighteen million years ago. This was based on the association of living apes with *Dryopithecus* and other apparent fossil apes, some of which were about eighteen million years old. Furthermore, there was another ancient form, *Ramapithecus,* known especially well from deposits in India and East Africa, that was supposed to be ancestral to everything that was considered hominid. The most compelling evidence was in the nature of the teeth of *Ramapithecus* and the subsequent hominids. Thus, *Australopithecus* and *Homo* were supposed to have their evolutionary roots in *Ramapithecus,* and *Ramapithecus* was found in some deposits as old as fourteen million years. If the interpretation of the fossil evidence was correct, the human and ape lineages had been separated for many millions of years.

Sarich and Wilson maintained that the human lineage was nowhere near that old. Their estimates placed the divergence of humans and apes no later than five million years ago. So, what of *Ramapithecus* with its hominidlike jaws and teeth? Sarich would write that "one no longer has the option of considering a fossil specimen older than about eight million years as a hominid *no matter what it looks like.*"

In 1980 Peter Andrews of the BM(NH) and I. Tekkaya of the Turkish Paleontological Service published the first description of a facial fragment of a variety of *Ramapithecus*. This dentally hominid ancestor turned out to have the facial characteristics of an orangutan. Two years later, David Pilbeam of Harvard would astonish the paleontological world with an almost complete specimen of this presumed hominid ancestor from deposits in Pakistan. There was no doubt about it. *Ramapithecus* and *Ramapithecus*-like "hominids" were actually fossil orang-utans.

Suddenly there was nothing that was hominid or hominid-related older than Lucy and her fellow *Australopithecus afarensis,* and they were well within the five-million-year mark. Sarich and Wilson appeared to have been right all along. Since Pilbeam's find, there has been some small effort to rebut the loss of *Ramapithecus* and the loss of human antiquity, but essentially most paleontologists have hopped onto the molecular bandwagon.

What emerges from this apparent disarray is that the orang-utan figures prominently in discussion of hominid evolution. *Ramapithecus* was dissociated from hominids because, while it possessed hominidlike teeth, it had the facial details characteristic of the orang-utan. And hominids do not. However, orang-utans, like *Ramapithecus,* also possess features of their teeth that are otherwise characteristically hominid.

When known only from jaws and teeth, *Ramapithecus* was eventually accepted by paleontologists as an undoubted ancestral hominid. When the facial skeleton of *Ramapithecus* became known, it and its fossil kin were relegated to an evolutionarily more distant position with the orang-utan. Why wasn't the orang-utan moved closer to *Ramapithecus,* and thus close to *Australopithecus* and *Homo?* It is because the African apes, the chimpanzee and the gorilla, if not the chimpanzee alone, are supposed to be closest to us evolutionarily. This theory of relatedness seems so real that the very idea that the orang-utan could have anything broadly to do with hominids is inconceivable. If such a theory of human—African ape relatedness is so strong, there must be ample evidence to support it. But is there?

3

The Search for
Our Closest Relatives

I DOUBT that many people have mentioned Jane Goodall, Clint Eastwood, and Bo Derek in the same breath. But in terms of exposure, Clint Eastwood and Bo Derek, through some of their movies, have almost been the Jane Goodalls of the orang-utan. Curiously, it is through these Hollywood movies that we get a sense of the orang-utan's being other than our poor evolutionary cousin. However, it is the chimpanzee and sometimes the gorilla who consistently get star billing in documentaries about human origins. In the popular consciousness, the apes are broken up and scattered evolutionarily away from humans: first the chimpanzee, then the gorilla, and last the seemingly least human of the humanlike apes, the orang-utan.

It was not always this way. In fact, the longest-lived proposal of relatedness among the hominoids was that the three large apes together constituted a real group of related animals. And it was to this great ape collective that humans were ultimately related. Only during the past twenty years or so has the taxonomic opinion of most systematists shifted to an alternative scheme that broke up the great ape family and allied humans with the African apes, leaving the orang-utan by itself.

In 1863, just four years after the publication of Darwin's *On the Origin of Species,* the young, self-assured anatomist Thomas Huxley published three essays under the collective title *Man's Place in Nature.* The first essay was "On the Natural History of the Man-like Apes"; it was a brief review of historical references to, and conceptions about, the apes. The third essay, "On Some Fossil Remains of Man," dwelt primarily on the somewhat modern-looking fossilized Engis skull from Belgium, but Huxley also provided some detailed discussion of the Neanderthal skull cap. However, it was the second essay, "On the Relations of Man to the Lower Animals," that had the most impact on the history of evolutionary studies. In this essay Huxley presented, for the first time, a broadly based and compelling morphological argument for recognizing humans as the animals, and the primates, that they are. If a dog or a monkey was a mammal, so was a human. If a gorilla was a primate, and a highly advanced primate, so was a human.

In 1896, Huxley republished these three essays with two more he had written during the intervening years. Among other things that had happened during the interval was the publication of Darwin's *Descent of Man*. And it was in no small way because of Huxley's argument of the reality of "Man's place in nature" that Darwin was able to expound at great length on the subject.

In the preface to the collected essays of 1896, Huxley reflected upon the history of his involvement in this study, which began in 1854, when he found himself having to teach "the principles of biological science with especial reference to paleontology." It was then that Huxley discovered his "lamentable ignorance in respect of many parts of the vast field of knowledge" that he was supposed to teach to others.

Of course, Huxley would inevitably run headlong into the issue of the relations of humans to other animals as well as to the positions on this issue of some of the most influential scholars. Among the British there was Charles Lyell, a powerful and important figure in Huxley's own scientific community. Lyell, the most eminent geologist of the day, is now considered to have been the father of modern geology. And until his convictions were eventually weakened by the constant irritations of Huxley and Darwin, Lyell was one of the most vocal of anti-evolutionists.

For Huxley "the position of the human species in zoological classification" was a serious problem:

> Even among those who considered man from the point of view, not of vulgar prejudice, but of science, opinions lay poles asunder. Linnaeus had taken one view, Cuvier another; and, among my senior contemporaries, men like Lyell, regarded by many as revolutionaries of the deepest dye, were strongly opposed to anything which tended to break down the barrier between man and the rest of the animal world.

The issue of "Man's place in nature" was such a searing question that "those who touched it," forewarned Huxley, "were almost certain to burn their finger severely."

In 1857 Sir Richard Owen, a powerful figure in British zoological circles, presented a paper at a session of the Linnaean Society which provoked Huxley into delving more deeply than he had intended into the structural differences and similarities between humans and apes. Owen's paper, "The Characters, Principles of Division and Primary Groups of the Class Mammalia," presented the argument that *Homo* could be separated from all other mammals on the basis of anatomical features, especially those of the brain, that were supposed to be unique to our genus. And because, in their brains, humans were supposed to be so superior to all other mammals, Owen proposed they should be set apart in their own group, to be called "Archencephala," those with the highest or most developed brains. Huxley's own investigations did not support these majestic conclusions. As Huxley believed then and would continually rediscover, one could not justify separating "Man" from the apes and monkeys, much less from other mammals.

It was at this point, in 1859, that *On the Origin of Species* appeared in print. Huxley took heart from Darwin's conclusion that "light will be thrown on the origin of man and his history" because Huxley's own investigations had been building up the steam necessary to generate that light. Darwin had opened the door of evolution more widely than it had ever been opened before, but he had left the human skeletons in the closet untouched. This was where Huxley came in. For, "inasmuch as Development and Vertebrate Anatomy were not among Mr. Darwin's many specialties," Huxley wrote, "it appeared to me that I should not be intruding on the ground he had made his own, if I discussed this part of the general question . . . In fact, I thought I might probably serve the cause of evolution by doing so."

The following year, Huxley brought his ideas on the relations of humans to other animals to the public through a series of lectures. To no one's surprise, Huxley's presentations provoked heated debate throughout the academic community as well as society. However, two years later, at a meeting of the British Association held at Cambridge University, Huxley found that he was not entirely alone. He had gained the support of the respected comparative anatomist Sir William Henry Flower, who demonstrated at those meetings that the supposedly unique features of the brain of *Homo sapiens* were also to be found in the brains of apes. In turn, in a joyous nineteenth-century burst of Latin, Huxley proclaimed, "Magna est veritas et praevalebit!" — truth is great and will prevail. But the way Huxley paraphrased it in translation was, "Truth is great certainly, but, considering her greatness, it is curious what a long time she is apt to take about prevailing."

By today's standards, and from the vantage point of what we now know, Huxley's argument for "Man's place in nature" is pretty slender. But when you consider the state of affairs of the mid-nineteenth century — not only of the scientific but of the political, social, and religious arenas — Huxley's presentation, only eighty pages long, is profound.

Perhaps the most striking aspect of Huxley's writing is that it is eminently understandable. He wrote for everyone.

> The question of questions for mankind — the problem which underlies all others, and is more deeply interesting than any other — is the ascertainment of the place which Man occupies in nature and of his relations to the universe of things.

This is a broad and unthreatening beginning. It deals with where and how we see ourselves with regard to everything else. Inevitably, Huxley points out, "the question of questions for mankind" will lead to "an inquiry into the nature and the closeness of the ties which connect him" not just with the universe at large, but with a small segment of it — "those singular creatures," the apes.

Huxley's argument begins with a discussion of the fundamental aspects of the development of the fertilized egg in the dog, from its original state to the eventual differentiation of the fetus and its placental link to the womb. Against this backdrop of fetal development Huxley then compares fetal development in the lizard, snake, frog, fish, bird, cat, opossum, and monkey. And we learn that "the embryos of a Snake and of a Lizard remain like one another longer than do those of a Snake and of a Bird; and the embryo of a Dog and of a Cat remain like one another for a far longer period than do those of a Dog and a Bird; or of a Dog and an Opossum; or even those of a Dog and a Monkey." Obviously, those organisms which remain like one another for a longer period of their embryonic and fetal development will emerge as more specifically the same type of animal. This follows even if the shapes of the animals at the end of development do not necessarily look exactly like each other, as is the case with a limbless snake and a typical lizard.

Embryology, the study of the earliest developmental phases of an organism, grew out of the general interest in the "great chain of being." As the Harvard evolutionary biologist Stephen Jay Gould has pointed out, the study of embryonic development would seem to be a natural extension of the practice of arranging adult organisms in an ascending ladder of life. Through the study of embryos and fetuses, it seemed that an organism became what it was going to become only after passing through the developmental levels of organisms anatomically less complex than itself.

The developmental history of an organism — ontogeny — was perceived as paralleling the arrangement of the organisms in the great chain. And the study of the ontogeny of an organism, from fertilized egg to embryo to fetus and beyond, could provide what the straightforward study of a string of adult organisms could not: it could provide a continuity between the different hierarchical levels of organisms. The comparative study of adults only left gaps in the chain. There was so much eagerness to demonstrate evidence of the great chain of being in the ontogeny of organisms that even aborted, poorly developed, and pathologic fetuses were thought to represent the "essences" of phases or links that belonged in the sequence of life's forms. A misshaped fetus could, for example, and with a bit of self-deception, provide proof of the crab phase "known" to be part of human development. The basic belief of these early embryologists was that each ontogenetic stage of an organism represented the *adult* individual of the organisms leading up to it. Thus, for example, the "gill slits" that one observes for a time in a human embryo were supposed to reflect the "fish phase" of human development.

Ernst Haeckel inherited these late eighteenth- and early nineteenth-century ideas and translated them into an evolutionary framework. Through the study of the ontogeny of an organism, he suggested, we can trace its evolutionary past. Haeckel referred to this as the Biogenetic Law, but it is more commonly spelled out as "ontogeny recapitulates phylogeny."

If evolution is seen in this light, there is only one practical way to achieve major evolutionary change: by adding a novel feature at the end of an ontogenetic sequence. It sounds simple enough — just tack one more developmental unit onto the continuum — but it means

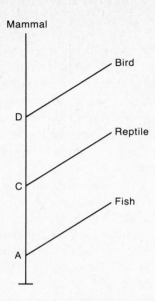

FIGURE 3 A   Karl Ernst von Baer's perception of ontogeny: organisms share
the general juvenile phases of development and then diverge along a more specific
course of development to emerge as the adult form, with its particularities, of a
fish, a reptile, a bird, or a mammal, for instance.

that only the adult individual will be seen as different. And we know,
even if only intuitively, that this is not the way evolutionary change
occurs. The twentieth century has had little use for the Biogenetic
Law.

Haeckel relied on the prevalent embryological dogma of the early
nineteenth century for principles that he could cast in an evolutionary
context. Huxley, on the other hand, drew on a less popular approach
to understanding an organism's developmental stages. And these ideas
stemmed from Karl Ernst von Baer.

Von Baer was not an evolutionist, and neither were his early nineteenth-century contemporaries in embryology. But von Baer's interpretation of the successive phases of an organism's development had
profoundly wide-reaching effects. He argued that the developmental
stages shared by organisms represented not the adults of lower links
but the *embryonic* or *fetal* phases of the organisms "left behind." For
example, humans and fish parallel each other in development by going

through a gill slit stage. At this stage, the human is not a fish, certainly not an adult fish, but neither is the fish at that stage a fully formed, adult fish. At that developmental point, however, the two animals will deviate from their common ontogenetic course. The fetal fish will develop its own set of features, culminating in those of adulthood, and the human will carry on in its own particular developmental path to become its own distinctive brand of organism.

Von Baer collapsed his ideas into four statements, which have come to be known as laws. The first law observes that during the development of an organism, general anatomical features appear before the more specialized ones. The second states that it is from the general characteristics that the specialized features develop. The third law notes that during an organism's development it becomes increasingly different from other animals. And finally, and most important, the early ontogenetic stages of an organism are *not* equivalent to the adult stages of those organisms lower in the hierarchy. Instead, the early ontogenetic stages are representative of the early ontogenetic stages of those less complex organisms.

Huxley used the logic of von Baer's laws to lead his reader to a particular conclusion. First, he made the point that "the more closely any animals resemble one another in adult structure, the longer and the more intimately do their embryos resemble one another." This would have to be true, since the more specialized features specific to any particular type of animal would appear later in the ontogenetic sequence. Thus, for example, it is not surprising to discover that the embryos of a dog and a cat remain more similar to each other over a longer period of time than do the embryos of a dog and a bird, or a dog and a monkey. This appears to be merely a matter of common sense. But if you accept this concept, you must also accept the ultimate conclusion. And thus Huxley can begin to close the trap:

> The study of development affords a clear test of closeness of structural affinity, and one turns with impatience to inquire what results are yielded by the study of the development of Man. Is he something apart? Does he originate in a totally different way from Dog, Bird, Frog, and Fish, thus justifying those who assert him to have no place in nature and no real affinity with the lower world of animal life?

The answer, of course, is a resounding "No!" If dogs, birds, frogs, and fish are united by a common ontogeny, then humans must be included as well. And if length of similar ontogenies reflects the ac-

quisition of similar anatomical details, then not only are humans
firmly entrenched in the animal world, they are most intimately associated with only a few members of that world, the apes. The study
of the development of the human fetus, from fertilized egg through
embryonic differentiation and the development of the placenta, led
Huxley to conclude that "without a doubt, in these respects, he is far
nearer the Apes, than the Apes are to the Dog."

Huxley was among the first to perceive that there were things
evolutionarily significant about the mammalian placenta, the nexus
of tissue and blood vessel that forms the vital connection between the
developing fetus and its mother. And it was in the configuration of
the placenta that Huxley found convincing similarity between apes
and humans, a similarity not found between humans and the other
mammal of comparison, the dog. Granted, there were definitely certain similarities in human and dog placental growth, especially in the
early phases of "yelk" (yolk) sac and allantois development, that
struck Huxley as "sufficient to place beyond all doubt the structural
unity of man with the rest of the animal world." But the more specific
similarities between humans and apes, such as "a spheroidal yelk-sac
and a discoidal, sometimes partially lobed, placenta," which would
appear later in their ontogenies, were irrefutable demonstrations of
the unity of humans "more particularly and closely with the apes."

Having thus established the unity of humans and apes, Huxley
provides other examples in support of this apparent "fact," and he
does it in a very interesting way. First, Huxley focuses on the two
African apes, proclaiming without any prior justification that "it is
quite certain that the Ape which most nearly approaches man, in the
totality of its organisation, is either the Chimpanzee or the Gorilla."
And then, of the African apes, he chooses to compare humans especially with the gorilla, stating rather matter-of-factly that "it makes
not practical difference, for the purposes of my present argument,
which is selected for comparison." I cannot entirely believe this, however. It seems to me that the innocence of Huxley's choice of the
gorilla as the model of comparison is betrayed by his reference to the
animal "as a brute now so celebrated in prose and verse, that all must
have heard of him, and have formed some conception of his appearance." As we shall see, it was fundamental to Huxley's argument that
the gorilla was the primary source of comparison.

While he is comparing humans and gorillas in order to discover the ways in which they are similar and those in which they are different, he will be comparing the gorilla with other animals, especially the monkeys, to determine the degrees to which these primates are similar or different. When all of these comparisons are completed, Huxley tells us, he can "inquire into the value and magnitude of these differences" and calculate where the greater morphological breach between these species occurs and on which side of it humans actually stand.

In carrying out his plan, Huxley compared humans with gorillas and gorillas with monkeys in the extent to which their limbs were either long or short, the configuration and details of their vertebral columns, the shapes of their pelvises, the anatomy of their hands and feet, and the number and general morphological disposition of their teeth, among other things. When there were differences to be found between humans and gorillas, there were even greater differences between gorillas and monkeys in the same anatomical region. Humans, obviously, stood closer to gorillas, and both were set distinctly apart from the "lower" primates.

As Huxley pointed out, the gorilla *looks* more like a human than does a chimpanzee, orang-utan, or gibbon, especially in the overall proportions of its arms and legs and of its hands and feet. For example, a gorilla's thumb is long relative to its finger length, just as our thumb is. In contrast, the thumb of a chimpanzee and especially of an orang-utan or gibbon is relatively quite short. The impression that these animals have a minuscule thumb is heightened by their having more elongate fingers as well.

The big toe of a gorilla is also relatively long. It is situated forward on the foot, near the other toes, and is more closely approximated to, or in line with, these other toes than it is divergent from them. Humans have a long big toe that is emplaced forward on the foot and very much in line with the other toes. In the chimpanzee, and especially in the orang-utan and in the gibbon, the "big" toe is appreciably smaller and situated much farther back and away from the other toes. In these primates, the big toe, when present (which it may not be in some orang-utans), is quite divergent from the other toes, giving the appearance of being rather thumblike.

The ways in which limb, torso, hand, and foot dimensions can be compared are seemingly endless and quickly become tedious. But what they do demonstrate is that in virtually all of its body proportions, the gorilla is the most similar among the hominoids to humans.

When Huxley came to discussing the brain, the chimpanzee and the orang-utan had to stand in for the gorilla because the gorilla's cerebral anatomy was not as well known as the other two apes'. But the brains of the chimpanzee and the orang-utan were more than adequate in providing Huxley with another resemblance to humans. Indeed, Huxley found that the brains of the chimpanzee and the orang-utan "fall but little below Man." So similar did Huxley perceive the brains of humans and apes to be that those differences he did find between these primates were relegated to the more banal parts of the body, such as the teeth, the pelvis, and the elements of the lower limb. Perhaps this was a reaction on Huxley's part to the weight given to the human brain, and especially to its much celebrated capacity for cerebration, in others' attempts to argue for the separateness of humans from the rest of the animal kingdom.

Given Huxley's presentation, one would have expected him to conclude that humans were most closely related to the gorilla, for it would seem that he chose this ape precisely because it would provide the most consistently positive comparisons to humans. But it probably wouldn't have been a surprise either if Huxley had suggested that humans and both of the African apes constituted a group. However, without any ado, Huxley ended up quietly grouping the "Man-like apes" — the chimpanzee, the gorilla, and the orang-utan — and with this group he associated "Man," "inasmuch as he differs less from them than they do from other families of the same order."

If you took to comparing humans and gorillas, then, not only would close resemblances be forthcoming, but they would be appreciated readily and easily by a potentially skeptical audience. A further consequence of the degree of overall similarity between humans and gorillas is that dissimilarities that could be demonstrated between a gorilla and a monkey, for example, would also lend themselves to being appreciated as dissimilarities between the reader and the monkey. By constantly making the statement that the degree of dissimilarity between humans and the apes, as exemplified by the gorilla, is less than it is between an ape and any lower primate, Huxley drew the apes in as a group — and as a group to which humans are most similar among primates, but from which humans are also set apart.

Adolph Schultz, a Swiss comparative anatomist and primatologist, took this form of reasoning one step farther. Starting in the 1920s, he amassed a huge amount of data demonstrating similarity between humans and one or another of the apes. But regardless of any special similarity shared by humans and a particular ape, Schultz maintained adamantly that the great apes were more closely related to each other than any one of them was to *Homo sapiens*. Schultz's lifework was really a defense of Huxley's more restrained grouping of the great apes as an assemblage to which humans were somehow related. And he was a formidable defender, opposing any and all other schemes of relatedness among apes and humans, especially those favored by some of the world's leading figures in primate and human evolutionary studies.

Of the various competing theories of human–ape relatedness proposed during the early decades of the twentieth century, Schultz seems to have been particularly threatened by the position taken by the German anatomist H. Weinert. Weinert had suggested that not only were chimpanzees and humans the most closely related of the hominoids, but the divergence between the human and chimpanzee lineages occurred relatively recently, as recently as, perhaps, the last five or so million years. In suggesting such a short period of separation between the human and chimpanzee lineages, Weinert was highly provocative. Regardless of disagreements over the specifics of relatedness of humans and apes, virtually all other primate systematists were committed to the idea that these primate lineages were quite ancient, perhaps even as old as forty or more million years. Such a long period of time seemed necessary for the human lineage to evolve its uniquenesses.

Weinert bucked the more popular of the competing arrangements of hominoid relationships by suggesting that there was a simple branching pattern. Gibbons split off first; then came orang-utans, gorillas, and finally humans and chimpanzees. On top of everything else, Weinert claimed that this last split occurred within the most recent geological past.

The interesting thing about Weinert's proposed phylogeny is that the sequence of divergence among the hominoids, and even the notion of a very late origin of the human lineage, is the same as the scheme proposed most recently by various molecular systematists. The similarity in interpretation between Weinert and these molecular systematists continues to the detail of concluding that humans and chimpanzees are virtually identical organisms.

The notion that humans and chimpanzees are the most closely related among the hominoid primates was accepted by a few scholars besides Weinert. However, the majority of systematists, regardless of differences in interpretation they might otherwise have, were overwhelmingly convinced of the unity of the African apes, the chimpanzee and the gorilla. These two apes were considered the closest of relatives — indeed, evolutionary sisters — in large part because they are so distinctive in anatomical features that are related specifically to their unique form of locomotion, knuckle walking. Another point on which most systematists agreed was the position of the gibbon relative to the other hominoids It seemed unquestionable that this primate had been the first, and thus the earliest, of the hominoids to branch off.

The powerhouse of the competing theories of hominoid relationships was the Huxley and then Schultz hypothesis of a great ape group, within which the African apes were the most closely related. Somewhere in the distant past, humans had shared a common ancestor with this great ape assemblage, but diverged long ago to forge ahead on their own evolutionary course. But even prior to this phylogenetic event, the lowliest hominoid, the gibbon, had been expelled from the major path of hominoid evolution.

The second most popular scheme of hominoid phylogeny maintained the African apes as sisters but regarded this group as the sister of the human lineage. The resultant branching pattern portrayed the gibbon as the first to diverge, then the orang-utan, then humans, and last the chimpanzee and the gorilla. A sometime forceful advocate of the special relatedness of humans to the African apes was the American paleontologist William King Gregory.

Gregory differed from most who came to work on problems of primate evolution and systematics in that he had been trained broadly in vertebrate paleontology. Even as his interests became increasingly focused on matters of primate and particularly hominoid evolution, Gregory saw this as a continuum in his work not only on mammals but also on fish and reptiles. One of his many tomes was even entitled *Our Face from Fish to Man*. His breadth was impressive, and his articles and monographs had a special ring of authority. But from what I can piece together from his many publications, which were directed at a spectrum of audiences of differing interests and levels of expertise, Gregory vacillated between two very different positions.

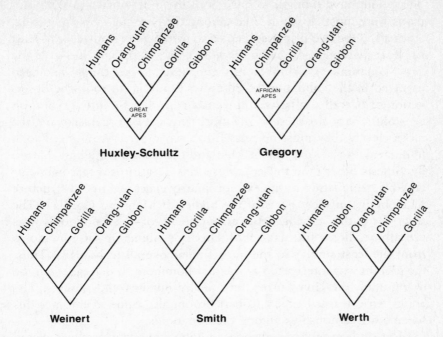

FIGURE 3B Five competing theories of the specific relationships of the living hominoids.

One was that humans constituted a group with the African apes. The other was that the great apes — the chimpanzee, the gorilla, and the orang-utan — were actually a real assemblage, but that humans' greatest similarities were to be found with the two African apes.

There were other suggestions of who was related to whom among the hominoids, but they, like Weinert's notion of a human–chimpanzee sister group, were attacked or dismissed as idiosyncratic or just plain wrongheaded. For example, there was the suggestion of the German paleontologist E. Werth that although the great apes did indeed constitute a group, this group was quite separate from humans, whose closest relative was in fact the gibbon. The role of the gibbon in human evolution had earlier been emphasized by Eugène Dubois, who came to believe that his fossil hominid from Java, *Pithecanthropus*, had evolved from a gibbonlike ancestor.

Another theory of hominoid phylogeny proposed the scheme that one would have thought to have been the most obvious — that humans were most closely related among the hominoids to the gorilla. After all, these two primates seemed to be the most similar, or at least the least dissimilar, of the hominoids. This was the conclusion of the British anatomist G. Elliot Smith, who, among other things, had been involved in the validation of Piltdown Man. Smith took the observation of overall similarity and translated it directly into a reflection of evolutionary closeness — the more similar two organisms are, the more closely they must be related to each other: "Any one who is familiar with the anatomy of Man and the Apes must admit that no hypothesis other than those of close kinship affords a reasonable or credible explanation of the extraordinarily exact identity of structure that obtains in most parts of the bodies of Man and Gorilla." The rest of Smith's phylogenetic arrangement of the hominoids was actually a direct translation of Huxley's lining up of hominoids from the least similar to the most similar overall to humans. Thus, the gibbon was interpreted as the first hominoid to diverge, then the orang-utan, the chimpanzee, and finally humans and gorillas. The South African fossil expert Robert Broom also came to embrace this scheme of relationships among the hominoids.

Most of these proposed theories of hominoid relatedness were based solely on comparisons between the living forms, following Huxley's approach. Fossils were typically not incorporated into the analysis, nor did they play a prominent role in the ultimate interpretation of phylogenetic relationships among the living forms. It does seem reasonable that one can learn more about the growth, development, and indeed the whole biology of organisms from the study of living forms than from the study of fragmentary fossils. Even Adolph Schultz didn't pay much attention to fossil material or to the interpretations of paleontologists, despite the fact that his work on the relationships among the extant hominoids spanned decades of intense field work and relevant paleontological discoveries.

Gregory, a paleontologist, did use fossils. In contrast to many other paleontologists, however, Gregory was quite well versed in the comparative skeletal and dental anatomy of living forms. And he was also not averse to using soft-tissue anatomy in his analyses of relatedness. Somewhat surprisingly, especially given the tendency during the past twenty years toward vicious debate between paleontologists and molecular systematists, Gregory was even impressed by the earlier studies of the British serologist George H. F. Nuttall. Nuttall had been among the first to try to demonstrate the evolutionary relationships of organisms by studying the immunological reactions between animals' blood proteins.

By the early 1920s, Gregory had taken up the defense of many theories on the specific relationships of extant hominoids to purported fossil relatives. But even though the publication of Huxley's essays and Darwin's *Descent of Man* were distant events, Gregory still felt compelled to argue the relatedness of humans to other hominoids in general before he could proceed to the details of hominoid evolution that he found more interesting.

Although Gregory favored grouping humans with the African apes, there was only one feature cited by him that could be demonstrated as being shared uniquely by humans and the African apes, a feature that continues to be cited as evidence of the evolutionary closeness of these three hominoids. Humans, chimpanzees, and gorillas are distinguished among primates by the development of frontal sinuses. These are air spaces in the region of the frontal bone just behind each eyebrow. Most of us are aware of the existence of our frontal sinuses only during allergy seasons or when under attack from some mighty cold or flu, when the membranes lining the inner walls of these sinuses flare up and become painful.

Gregory was for years an idol of mine, even though I came to disagree in print with some of his interpretations of various very early fossil primates. But my review of Gregory's writings on the details of the relationships among the hominoids caught me by surprise. I had not expected to find a passage like this one in Gregory's argument for the relatedness of humans to the African apes:

> The anatomy of the nose testifies in the same direction. According to Professor J. Howard McGregor the gorilla nose, although repulsive in human eyes, has in it all the "makings" of a human nose, and acquires chiefly a forward and downward growth of its tip to be transformed into a subhuman type. The lowest existing types of human nose *(fig. 302)* indeed retain much of the gorilloid heritage.

The caption of text figure 302 reads, "A Tasmanian Man, Showing a Very Low Type of Face, with an Excessively Wide Flat Nose." Obviously, Tasmanians were not considered with much more regard than eighteenth-century taxonomists had for the Hottentot, that supposed human "link" to the subhuman primates. And Gregory continues:

> The eyes of the great apes could hardly be more human than they are.
> The fundus of the eye of a chimpanzee, as figured by Johnson (1901),
> exhibits the most detailed resemblance to that of a negro.

Truly, our perceptions about the world can greatly color our interpretations of it. It seems that the idea of the "great chain of being" was all-pervasive. Religious adherents to this notion sought to prove God's, and, no doubt, their own, greatness by seeking out the representative of every link in that hierarchically ascending chain. Nineteenth-century evolutionists gave it another life and another meaning, but the concept of a progression of life's forms, from the simplest to the most complex, with the continuum carrying through a ranked ordering of humans, lived on. Darwin used it in his argument for the rise of civilized "Man." And its influence invaded the twentieth century, seemingly providing the proof of relatedness that the study of teeth, bones, and anatomy could only hint at. One might well wonder what the "great chain of being" and its evolutionary successor would have looked like had its inventors been gorillas, orang-utans, Hottentots, or Tasmanians.

Perhaps it is fortunate that noses, eyes, skin, and other soft-tissue bits and pieces do not fossilize. It may be a nuisance to have to deal with typically fragmentary jaws or bones of now extinct animals, but in the absence of the completeness of the individual, a paleontologist is forced to consider the detail of tooth or bone morphology. However, finding a piece of mineralized bone or tooth in the "right" place and in the "right" deposit can also provoke a substantial amount of speculation on its own.

A seemingly straightforward way in which to go about "discovering" the affinities of the extant hominoids is to try to tie them to specific fossils. In this way, one might begin to piece together a temporal picture of the history of the members of the group. And when one begins to talk about extant hominoids and potential fossil allies, one cannot proceed for very long without referring to *Dryopithecus*.

By the 1920s, there was some support for the broad view that *Dryopithecus,* and other Miocene *Dryopithecus*-like forms, were somehow integral to the subsequent evolution of hominids and apes.

Also by this time, discoveries had been made of Miocene "apes" in areas other than in Europe. Most of us think of East Africa as the home of these sites of Miocene "apes." However, well before any East African deposit had been successfully prospected for fossil hominoids, important specimens had been found in deposits of the Siwalik Hills of the northeastern part of India.

The first specimen of any significance to come from the Siwaliks was discovered in the 1870s by the British paleontologist Richard Lydekker. The specimen consisted of a partial left upper jaw with almost all of its teeth preserved. Lydekker proposed the genus and species names *Paleopithecus sivalensis* for this new hominoid. Some sixty years later, G. Edward Lewis, then a Yale graduate student working on the Siwalik hominoid material for his dissertation, pointed out that the genus name *Paleopithecus* had already been used to refer to the extinct reptile that had left a series of fossil footprints. Thus, according to the rules of zoological nomenclature, the genus *Paleopithecus* could not be used to refer to Lydekker's hominoid. Lewis then transferred the specimens of the species *sivalensis* to a different genus, *Sivapithecus*. And it is *Sivapithecus* that, along with *Ramapithecus,* will emerge as more central than anyone would have imagined to the issue of human origins and to the question of who our closest living relative is.

Lydekker continued to collect specimens from Siwalik deposits. Among them was an upper canine that is essentially indistinguishable from the upper canine of a modern orang-utan. However, it was through the efforts of another British paleontologist, Guy Pilgrim, who had been attached to the Geological Survey of India, that fossils really were amassed. Between 1910 and 1915, Pilgrim found and described fossils that would become the type specimens of many new genera and species.

Pilgrim thought he had found new species of the otherwise European genus *Dryopithecus*. One of these he named *Dryopithecus punjabicus*. The type material of *D. punjabicus* consisted of two mandibular fragments thought to have come from the same individual. Pilgrim suggested that this new species of fossil hominoid was related to the gorilla. Another species of *Dryopithecus, D. giganteus,* Pilgrim saw as being possibly related to the chimpanzee. Pilgrim also thought that Lydekker's *Paleopithecus sivalensis* specimens looked like the result of a cross between a chimpanzee, a gorilla, and a gibbon.

One of the new genera that Lydekker proposed was *Paleosimia*. The type specimen of this genus was only one isolated upper right molar, but its wrinkled enamel surface became the subject of the species name, *rugosidens*. *Paleosimia rugosidens'* upper molar was reminiscent of an orang-utan's.

Another creation of Pilgrim's was the new genus and species *Sivapithecus indicus,* which was based on part of a lower jaw. The genus name derives from the name of the Indian god Siva. In keeping with his previous acts of deriving each of the extant large apes from a specific fossil species, Pilgrim made the even bolder move of suggesting that *S. indicus* had been near the ancestry of the human lineage. Pilgrim even went so far as to place *S. indicus* in our own taxonomic family, Hominidae.

William K. Gregory was quick to respond:

> The important genus and species *Sivapithecus indicus,* from the Lower and Middle Siwaliks, rest upon fragments of the lower jaw and dentition. From these Dr. Pilgrim has attempted a restoration of the lower jaw that shows a subhuman divergence of the opposite rami and a very short, manlike symphysis. Pilgrim regards this genus as in or near the ancestral line of *Homo sapiens.*
>
> The reviewer regrets to report that after a careful study of the evidence he believes Dr. Pilgrim has erred in attributing the above-mentioned human characteristics to *Sivapithecus,* the jaw of which, in the reviewer's opinion, should be restored rather after the pattern of the female orang jaw.

What was important for Gregory in discriminating between apes and hominids was the disposition of the canine tooth and the bicuspid teeth behind it, the premolars. Since *Sivapithecus indicus* had what Gregory considered "ape-like canines and front premolars," like an orang-utan, the fossil could not belong to Hominidae. Gregory placed the fossil genus in the family Simiidae, which was the taxonomic name then used to subsume the apes.

The essence of this exchange between Gregory and Pilgrim is summarized in the title of Gregory's 1915 publication, "Is *Sivapithecus* Pilgrim an Ancestor of Man?" It is, if Pilgrim is right. But if Gregory is correct in dissociating *Sivapithecus* from the hominids because of its apparently greater similarity to the orang-utan, then we could easily ask the question "Is *Sivapithecus* Pilgrim an ancestor of the orang-utan?" This is precisely the crux of the present-day debates on hominoid phylogeny and human origins.

In the 1930s, just a few years after Hopwood and his colleagues had begun to find specimens of *Proconsul* in East African deposits, G. Edward Lewis undertook the analysis of the fossil hominoid material from the Siwalik Hills of India. Almost immediately Lewis recognized two new genera, *Ramapithecus* and *Bramapithecus,* which he thought were closer to hominids than any other fossil genera or species then known.

The type specimen of *Ramapithecus* was a good-sized piece of a right maxilla, or upper jaw. Lewis's reconstruction of the probable length of the maxilla indicated that *Ramapithecus* had a short face, which is characteristic of modern humans. To reflect this feature, Lewis coined the species name *brevirostris.*

The new genus and species *Bramapithecus thorpei* was based on a partial left lower jaw which preserved its last two molars. Lewis felt that this specimen of *Bramapithecus* demonstrated both a closeness with *Dryopithecus* and a position possibly ancestral to the human lineage. Lewis also thought Pilgrim's *Dryopithecus punjabicus* was just another species of *Bramapithecus,* to which he transferred the species. So there were two species of *Bramapithecus,* the original *B. thorpei* and the referred *B. punjabicus.*

Barnum Brown, of the American Museum of Natural History in New York City, had also been collecting fossil hominoids from Siwalik deposits. As a result of the museum's activities, its most eminent paleontologist, William K. Gregory, became more involved in the study of hominoids. In 1938 he and Milo Hellman, who had quite a reputation in comparative dental studies, joined Lewis in describing various Siwalik specimens. Among other things, these three collaborators ended up assigning to the genus and species *Ramapithecus brevirostris* a fragment of a lower jaw that was apelike, not hominidlike. This unfortunate association of something obviously apelike with an obviously hominidlike specimen created such a muddied situation that further consideration of *Ramapithecus'* being related to hominids came to a halt until about 1960, when Elwyn Simons, then a young professor of vertebrate paleontology at Yale, "rediscovered" Lewis's hominid while rummaging through the dusty drawers of the fossil collections at the Peabody Museum of Natural History.

In 1961 Simons published a brief paper in which it was stated clearly for the first time that *Ramapithecus* should definitely be regarded as a true hominid. Simons based this assertion in large part

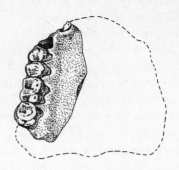

FIGURE 3C The type maxilla of *Ramapithecus,* discovered in the Siwalik Hills of India in the 1930s and thought by G. E. Lewis to be close to hominids in part because this fossil hominoid seemed to have a short face *(left)*. In the 1960s Elwyn Simons argued that *Ramapithecus* was a "good ancestor" of hominids in part because he thought the shape of the dental arcade was short, "parabolic," and broad posteriorly, as in humans *(right)*.

on his reconstruction of the shape of *Ramapithecus'* palate. In contrast to the long, squared-at-the-front, narrow, almost-parallel-sided palate of a chimpanzee or gorilla, *Ramapithecus* appeared to have a short and broadly curved palate, as is characteristic of members of the genus *Homo.* We now know that this reconstruction is inaccurate, but that doesn't adversely affect Simons's interpretation.

Years of debate left little doubt that *Ramapithecus* was definitely hominid in its teeth. *Ramapithecus* had premolars and molars, the cheek teeth, with low crowns and rather flat chewing surfaces. If you run your tongue over your three molars (assuming that you have all three), starting in the back of your mouth and then moving forward onto the two premolars, you will appreciate the low topography of these teeth. If you were a gorilla, the cusps (points) of these teeth would be more elevated and conical. But in addition to having low-cusped cheek teeth, *Ramapithecus* was found to have a thick layer of enamel covering its molar teeth, which is also characteristic of "proper" hominids. If you were to take a thin section through one of your molars, you would also see a thick cap of enamel covering and protecting the inner sensitive anatomy of the tooth. The molar enamel of a chimpanzee or gorilla is quite thin.

In 1962 Louis Leakey, who had participated in the collection and description of additional *Proconsul* material from East Africa, announced the discovery of a specimen of a different type of hominoid which he thought showed "a greater or lesser approach towards the structures we associate with the Hominidae." Although Leakey did find similarities between his new hominoid and the *Ramapithecus* material from India, he preferred to create a new genus and species to accommodate his hominidlike fossil maxillary fragment. Thus was born *Kenyapithecus wickeri,* named after its discovery at Fort Ternan, Kenya, a site originally found by Fred Wicker. The East African deposits that yielded *Kenyapithecus* were dated at approximately fourteen million years, whereas *Ramapithecus* had come from Siwalik deposits that were younger by two or more million years. This difference in geological age, as well as the fact that *Kenyapithecus* and *Ramapithecus* had been found in widely separated continents, were of primary importance in Leakey's maintaining a separate genus for his early hominid.

The following year, Simons reviewed the material on which the two genera *Kenyapithecus* and *Ramapithecus* had been based. He concluded that they were basically the same animal and suggested that Leakey's genus *Kenyapithecus* should be dropped, or, as taxonomists say, sunk. At the same time, Simons realized that Pilgrim's *Dryopithecus punjabicus* was actually a specimen of *Ramapithecus,* so he transferred the two mandibular fragments originally assigned to *Dryopithecus punjabicus* to the genus *Ramapithecus*. But because Pilgrim had proposed the species name *punjabicus* years before Lewis created the species *brevirostris,* Simons was obliged by the rules of zoological nomenclature to reconstitute *Ramapithecus* as *Ramapithecus punjabicus*.

A few years later, Simons reviewed Lewis's genus *Bramapithecus* and concluded that the mandibular pieces that had been referred to this genus were nothing more than additional representatives of *Ramapithecus punjabicus*. Thus, *Ramapithecus punjabicus* came to house most of the material that at one time or another had been thought to be in or near the ancestry of hominids.

In 1965 Simons and one of his graduate students, David Pilbeam, reviewed all of the Miocene hominoid material and concluded that apelike forms had been present throughout Europe, Africa, and India. Simons and Pilbeam felt that all these apelike hominoids were sufficiently similar to each other that keeping them in separate genera was unnecessary. They suggested that *Dryopithecus* should be the genus for all of these species, including the material that had been assigned to *Sivapithecus,* even though *Sivapithecus* was, in its cheek teeth, distinctly different from the others allocated to *Dryopithecus.* However, because of its large, apparently apelike canines and somewhat elongate lower first premolar, *Sivapithecus* came to join the East African *Proconsul* and the European *Dryopithecus* as a subordinate genus to the major genus *Dryopithecus.*

Dryopithecus thus became represented by a swarm of species of Miocene fossil apes from which it was then possible to find the ancestry of each of the modern great apes. Pilbeam even went so far as to suggest that a small form of the *Dryopithecus* subgenus *Proconsul* had given rise to the smaller of the African apes, the chimpanzee, and that a larger species of *Proconsul* had been ancestral to the gorilla. Farther to the east, *Sivapithecus* had given rise to the modern orang-utan as well as to a now extinct, enormous hominoid with the not unsurprising genus name *Gigantopithecus.* *Gigantopithecus* had been discovered when G. H. R. von Koenigswald found that its isolated teeth were being sold in Chinese drugstores as dragon teeth, which were ground up for medicinal use. A small species of *Gigantopithecus* is known from Miocene deposits in India, and a huge species — much larger than a gorilla — has been found in deposits in China that are only a few million years old.

In contrast to the fossil apes, which were represented by a triumvirate of subgenera and a horde of species, hominids could trace their apparent ancestry back only to the single-specied genus *Ramapithecus.* Since the human and ape lineages were supposed to have had a common ancestor in the distant evolutionary past, *Dryopithecus* was once again sought to provide this link. But while specific species of *Dryopithecus* subgenera had been pinpointed as ancestors of the modern apes, the progenitor of *Ramapithecus* was left purely in the realm of the hypothetical. However, the search for this ultimate ancestor continued, spurred on by Darwin's conviction that "some ancient member of the anthropomorphous subgroup gave birth to man."

By the late 1960s, the general outline of hominoid evolution had been agreed upon. The gibbon was the most primitive of the extant hominoids, and its earliest relatives were the Miocene *Pliopithecus* from Europe and *Limnopithecus* from East Africa. The ancestry of each of the extant great apes, which were thought to constitute a group, could be traced to a particular species of Miocene *Dryopithecus.* Species of the East African subgenus *Proconsul* had been ancestral

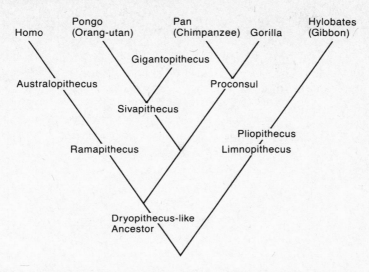

FIGURE 3D The commonly accepted scheme during the 1960s of relationships among fossil and living hominoids. One of the major features linking *Ramapithecus* with *Australopithecus* and *Homo* was its possession of low-cusped cheek teeth and thick molar enamel.

to the African apes, and a species of the Asian subgenus *Sivapithecus* had been ancestral to the Asian ape, the orang-utan. A species of *Sivapithecus* had given rise as well to the now extinct hominoid *Gigantopithecus,* which was also native to Asia. And some very primitive species of *Dryopithecus* yielded the earliest hominid, *Ramapithecus,* which in turn gave rise to *Australopithecus,* from which our genus, *Homo,* eventually evolved.

There were from time to time rather heated debates over the identification of *the* hominid ancestor. The favorite alternative to *Ramapithecus* was *Gigantopithecus,* which appeared to be even more hominidlike in that its cheek teeth were much lower-cusped and flatter. In addition, the canine teeth of *Gigantopithecus* were supposed to be relatively smaller and incorporated into an anterior dental cutting unit with the more chisellike incisors. In *Ramapithecus,* the canine was relatively taller and distinct from the incisor region. The argument became moot, however, when it was realized that one could unite *Ramapithecus* and *Gigantopithecus,* and *Sivapithecus* as well, with *Australopithecus* and *Homo* because all of these hominoids have

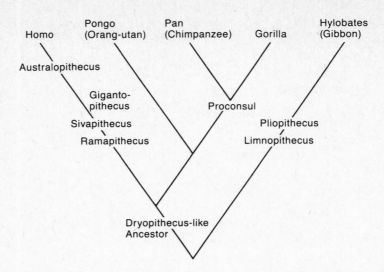

FIGURE 3E The scheme of relationships among fossil and living hominoids that replaced the one in figure 3D after *Gigantopithecus* and *Sivapithecus* were found to have low-cusped cheek teeth and thick molar enamel, features that would link them with *Ramapithecus* and, more broadly, *Australopithecus* and *Homo*.

low-cusped, flat-surfaced cheek teeth. In addition, these potential or "real" hominids were found to have thick enamel covering their molar teeth. Thus, *Ramapithecus, Gigantopithecus,* and *Sivapithecus* shared with *Australopithecus* and *Homo* those dental features that would consistently distinguish and delineate a particular group of hominoids.

It would seem to make sense that taxonomic names should subsume a collection of organisms that actually constitute a coherent evolutionary group. After all, if a collection of organisms can plausibly be united because of distinctive features they share, features that are not also found in other animals, then they might have a reality that could be referred to formally in a classification. Everyone agrees that rodents are an evolutionary group and that there are specific features — such as large, continually growing anterior gnawing teeth — that reflect the evolutionary unity of these animals. The group is formally recognized as the taxonomic order Rodentia. However, when one is dealing with our own evolutionary group, or potential candidates for inclusion in our own group this seemingly objective procedure becomes a highly sensitive issue.

If *Ramapithecus, Sivapithecus,* and *Gigantopithecus* are united ev-olutionarily with the "proper" hominids *Australopithecus* and *Homo,* then you might conclude that the five genera should be united tax-onomically as well. Thus, if there was a significant evolutionary re-lationship among any of these different genera and species, or taxa, then we could discuss the details of evolution within Hominidae. However, this move seemed too much for most paleoanthropologists to swallow, and the "un-proper" hominids were relegated to a sep-arate family, Ramapithecidae.

This was how the situation shaped up in the 1970s from the fossil side of things. From the molecular systematist's perspective, however, this was not the picture of hominoid evolutionary events at all.

Right from the start, it was obvious that there would be disagree-ment between a morphological, including paleontological, view of hominoid evolution and one provided by molecular studies. Molecular systematists never really supported the unity of a great ape group; instead, they concluded that the African apes were more closely related to humans than they were to the orang-utan. The general molecular picture of hominoid evolution portrayed the branching off first of the gibbon, then of the orang-utan, and then of the members of the human–African ape group.

Many of the molecular approaches to the problem had to stop at the level of asserting that humans and the African apes shared a common ancestor after gibbons and then orang-utans diverged. These studies could not resolve what appeared to be a trichotomous, equi-distant association of humans, chimpanzees, and gorillas. But other molecular systematists thought they could demonstrate that humans were the sister taxon of a chimpanzee–gorilla group. At least the recognition of a chimpanzee–gorilla sister group was seemingly con-sistent with morphology. However, occasional molecular reports hinted at a closer association of chimpanzees to humans than to gorillas.

Many morphologists and paleontologists went quietly along with the dissolution of the great ape group and accepted the theory that the African apes were more closely related to humans than to orang-utans. With a minimal shifting around of fossils, this general scheme could be accommodated. Everything could still be derived from species of *Dryopithecus,* so there was no real problem. However, when mo-lecular systematists started to use their data to calculate the dates at which the various hominoids diverged, and, in some instances, these dates turned out to be greatly at odds with those derived from fossils,

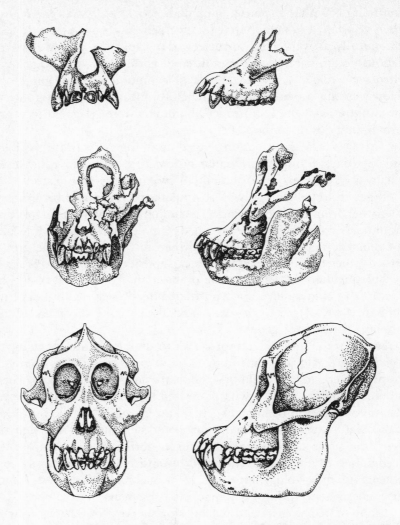

FIGURE 3F Remains of the orang-utan-like fossil *Sivapithecus* (*top*, from Turkey; *middle,* from Pakistan) compared with a skull of a modern orang-utan (*bottom*). Noticeable among the uniquenesses shared by both the fossil and the living hominoid are the "pinching" of the face behind the upper canine root, the smallness of the nasal aperture, the closeness of the highly positioned, oval, flattened orbits, the flatness of the cheek bones, and the protrusion of the front of the upper jaw.

the honeymoon of benign collaboration between molecule advocates and paleontologists came to an abrupt halt. The major focus of the debate and controversy was *Ramapithecus*.

With *Ramapithecus* dated in East Africa to fourteen million years, the human lineage must have diverged at an earlier time if *Ramapithecus,* or something *Ramapithecus*-like, was indeed a possible human ancestor. The diversification of the great apes, in whatever order one preferred, would also have been quite early, as indicated by the fact that some specimens of *Dryopithecus* were perhaps as old as eighteen million years.

According to the molecular systematists Vincent Sarich and Allan Wilson of Berkeley, however, these dates were much too early. Instead, Sarich and Wilson arrived at the conclusion that the divergence of humans and the African apes could not have occurred prior to five, or at most eight, million years ago. Thus, as Sarich would assert, *Ramapithecus* could not be a hominid, even if it looked like one. And Sarich was adamant about his results, especially the dates of divergence he had calculated.

But not all molecular systematists were in agreement with Sarich and Wilson. Another leading molecular systematist, Morris Goodman, tried to accommodate the dates of hominoid divergence derived from the fossil record. As a result of the ensuing debates, Sarich and Wilson found themselves at odds with just about everyone — paleontologists and molecular systematists alike. The 1970s saw an acceptance of the breakup of the great ape group, but a reliance on the fossil record for the historical elements of hominoid evolution persisted.

In 1980, the entire picture of hominoid, and especially hominid, evolution was dismantled as a result of the discovery, in Miocene deposits of Turkey, by Peter Andrews and I. Tekkaya of a large portion of the facial skeleton of *Sivapithecus*. First off, Andrews and Tekkaya argued that *Sivapithecus* was essentially the same animal as *Ramapithecus*. And since *Sivapithecus* had been named by Pilgrim prior to Lewis's erection of the genus *Ramapithecus,* the name *Sivapithecus* had precedence. But of more profound consequence than their reducing *Sivapithecus* and *Ramapithecus* to a single genus was the realization that the facial skeleton of *Sivapithecus* was distinctly orangutan in its features. This conclusion became even more evident when a more complete specimen of *Sivapithecus,* from Pakistan, was sub-

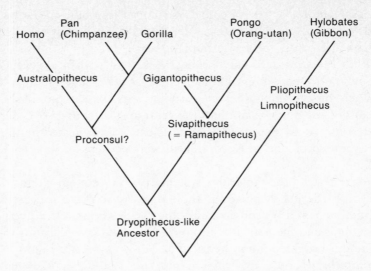

FIGURE 3G The recently popular scheme of relationships among fossil and living hominoids. Based in large part on the results of biomolecular studies, the great ape group is disbanded. Because *Sivapithecus* (with which *Ramapithecus* is now synonymized) has facial and dental details characteristic of the orang-utan, the notion of a thick-enameled hominid group is also jettisoned.

sequently found by David Pilbeam. There was no doubt about it. In its high, rounded orbits, the narrow space between the eyes, its flat face and cheek bones, and many other aspects of its facial, palatal, and dental anatomy, *Sivapithecus* shared with the orang-utan those features that are indelibly stamped "orang-utan."

A few primate paleontologists, most notably Elwyn Simons and his younger colleague at Duke University, Richard Kay, have not been convinced of the orang-utan likeness of *Sivapithecus*. However, most of us are convinced and accept the similarities between *Sivapithecus* and the orang-utan as evidence of their relatedness. Where I depart from all others is in the incorporation of this new information into a broader theory of relationships among the hominoids.

The common response to the situation is that if *Sivapithecus* is really a fossil orang-utan of sorts, then it cannot be a hominid. Thus, the general tendency among my colleagues has been to reject as being phylogenetically significant those dental features that had otherwise so successfully united *Sivapithecus* with the "proper" hominids. With the loss of *Sivapithecus* as a potential hominid ancestor, the earliest-known "real" hominid becomes *Australopithecus,* which at present is not known from deposits older than four million years. This would seem to vindicate Sarich and Wilson's calculation of a relatively recent divergence of the human lineage. The once species-abundant hominid group-at-large has been deflated to merely two genera, *Australopithecus* and *Homo,* while the orang-utan has inherited a ready-made fossil family.

There are two possibilities here: either *Sivapithecus* is related to hominids or it is related to orang-utans. My response is to wonder whether all three taxa — hominids, orang-utans, and *Sivapithecus* — are not somehow related. But this has not been a question that has struck many as even being something to consider. The reason for this neglect is that the *Sivapithecus* material has been interpreted within the context of a certain theory of hominoid phylogeny, a theory that maintains that one or both of the African apes are more closely related to humans than are orang-utans. With the assumption that the African apes intervene phylogenetically between humans and orang-utans, the features that might demonstrate the unity of hominids, *Sivapithecus,* and orang-utans are not given any evolutionary credibility.

Some of my colleagues think there is a lot of morphological support for uniting humans with the African apes. Gregory's work is among those cited, and Darwin is given the ultimate credit for defending this hypothesis. However, I find that I must remain much more skeptical. In my review of the literature, I have not been able to justify more than a handful of possibly evolutionarily significant features that would attest to a close relationship between humans and the African apes.

The present-day adherence to the association of humans and the African apes is based primarily on the conclusions of molecular and chromosomal studies, which frequently demonstrate that the gorilla, and especially the chimpanzee, are most similar overall to humans. However, the criterion for determining relatedness is, as it has been for studies based on morphology, overall similarity.

I believe there is something very wrong here. Overall similarity between organisms may not in every instance be an accurate reflection of their evolutionary closeness. We cannot pretend to know beforehand what the relationships might be or what features will be revealed as reflective of such relationships. Similarity means different things evolutionarily. Its demonstration is only the first small step toward unraveling the evolutionary relationships of organisms.

4

The Riddle of
Relatedness

YOU ARE A DETECTIVE on the Evolutionary Police Force, assigned
to the Special Phylogenetic Reconstruction Squad. Your job is to try
to figure out the mysterious links between the members of the Genus
and Species Gang. You've been on the case for months, but it seems
like years. You've scrounged around for clues in the badlands of East
Africa and sorted through the dusty collections and fading records in
the world's natural history museums.

"It's a tough job," you say to yourself, thinking back over all the
hardships in the field and all those meals in Paris that you've had to
endure, "but somebody has to do it."

The side door to the stage in the lineup room opens, cutting short
your reminiscences. You've been waiting for this moment for weeks.
All of the possible members of the Genus and Species Gang have been
rounded up for this, the biggest investigation in years. There are the
youngsters in the bunch — the chimps, who work in large groups,
the quiet gorillas, the solitary orangs, who like to work alone. And
there are the more elusive oldsters, the "fossils" of the Gang. Some
of them are incredibly old. They come from all over — Asia, Africa,
Europe. Sometimes you get a whole one, but usually you've had to
content yourself with a brief glimpse, a small clue as to their identity.

It's your turn to interrogate the suspects. You start in on *Homo
erectus.*

"Who got you into this in the first place? It was *Homo habilis,*
wasn't it? And *Homo habilis* got pushed into it by *Australopithecus.*
Now you're setting up Neanderthals, aren't you?"

No answer. You continue:

"Why did it happen? Was it your brains — swelled with evolutionary success? Maybe it was running faster? Making fire? Making speeches?"

Still no answer. You press on:

"So, when did this all happen? Two, five million years ago, sometime in between, sometime even earlier? How did it happen? Where did it happen? Where's the next event going to take place?"

You go on all night, asking your how, why, when, where questions of every single one of them. And in the end you put together a report explaining how the Genus and Species Gang got to be what it is, how the youngsters evolved from the older ones, and how these older ones evolved from the even older "fossils." There were a lot of evolutionary pressures to being in the Genus and Species Gang, lots of competition. That guy Darwin said it would all happen — at least the important parts — in Africa, and, by God, it certainly seems to have happened like he said it would. Look at all the fossils that hung out there — hominids and apes all over the place.

You've convinced yourself of how the Genus and Species Gang could have evolved and been so successful at it. Now you have to convince the Chief.

"I like this stuff about straight succession among the Species in the Genus *Homo* line," he says. "Bigger brains, better legs, more ingenious about making things, maybe some family and community life, articulate. Makes sense that they evolved this way. It's as plain as the nose on your face."

That was the end of the good news.

"This other stuff, though," he continues, "I don't like it. You'll have to do better. Some of those *Australopithecus* must be lying about their age — it's how old they are that counts, you know. Come back when you've figured better how and why the *Homo* genus would split off from the *Australopithecus* genus anyway. And which *Australopithecus* was responsible for it? No. This stuff just doesn't make any sense. Make it fit."

Much of what is written in the name of evolution is scenario, what Stephen Jay Gould has recently referred to as the "just so" story. And the questions that are typically raised — and answered — are, How, Why, When, and Where?

How did this species evolve from that one?

Why did it do so?

When did this happen?

And where?

The primary task is to piece together a solid evolutionary argument that will convince you, the audience, of the reality of the underlying scheme of ancestry and descent. But no matter how persuasive that argument, you will not have been treated to a rigorous discourse on the criteria that were used in the first place to generate that arrangement of relationships. The relationships are presented as a given; the real priority is to defend them. If I can come up with a more smoothly argued scenario than another on how and why, when and where, and from whom your species evolved, then my basic assumptions about their relatedness must be correct.

A recurrent debate in paleoanthropology has been the evolution of human bipedalism, walking. How did it evolve? Why? From what condition did it evolve? Why?

For a while, a popular scheme was that the gibbon represented the animal, or at least the type of animal, from which hominid bipedalism evolved. The gibbon is not just a good arborealist, it is the only true brachiator among primates, hanging and swinging from branches with its arms fully extended. Thus, the trunk of the gibbon is held upright and the animal's head is balanced atop this vertical strut. According to the scenario, the gibbonlike proto-hominid could merely have dropped to the ground and transported itself on two legs. Above the waist, this proto-hominid would already have been "preadapted" to carrying its torso upright.

However, another scenario was proposed around the turn of this century which argued for the derivation of the human lineage from the small, enigmatic southeast Asian primate *Tarsius*, the tarsier. The tarsier has in one way or another been the center of intense debate and controversy since then. Is it more closely related to the "lower" primates, the lemurs and lorises? Or is it more intimately related to the "higher" (anthropoid) primates, the New World monkeys, the Old World monkeys, and the hominoids?

One reason for thinking that *Tarsius* might have something to do evolutionarily with anthropoids is that it has a short snout, rather large eyes, and a rounded, globular cranium. In its overall craniofacial appearance, the tarsier looks like many of the "higher" primates. A possible reason for linking the tarsier rather directly with the human

FIGURE 4A   *(Left)* The tarsier *(Tarsius)* clinging to a vertical support. Note, for example, its two pedal grooming claws, relatively hairless tail, exceedingly large eyes, and elongate foot.

(Middle and right) The dark and light phases of the white-cheeked or white-handed gibbon *(Hylobates lar)*. Its elongate arms enable the gibbon to brachiate, or arm-swing, but they get in the way when the animal is on the ground. In the suspensory locomotion of the gibbon and the vertical clinging of the tarsier, the torso is held upright and the head is balanced atop the vertebral column — as in bipedal humans.

lineage is that the tarsier balances its "cute" little head on a vertically held vertebral column. This posture in *Tarsius* is a consequence of the animal's tendency to perch clutching vertical branches and reedlike plants. When bounding between vertical supports or pouncing upon some small victim scurrying through the forest floor débris, *Tarsius* is bipedal, but only in the sense that the kangaroo rat, for example, is also bipedal. The evolutionary chasm between the erect posture of a tarsier and the erect bipedalism of a human was not too broad for some paleontologists' leap of faith. And if humans and *Tarsius* are indeed related, these paleontologists argued, then their common ancestor existed well over forty-five million years ago, in Eocene deposits, in the form of a tiny, tarsierlike fossil called *Anaptomorphus*.

The tarsier model for the evolution of human bipedalism, although attracting some rather powerful supporters, was greeted with a flurry of rebuttal. One of the most outraged opponents of this scheme was William K. Gregory, who argued that in order to accept such antiquity for the human lineage, one "must postulate the existence of a long series of genera and species representing an unknown and very distinct family of primates, ranging from the Lower Eocene onward." Thus, there should be a very visible fossil record of this long string of ancestors throughout those tens of millions of years. However, as Gregory was quick to point out, "no trace has been found in any age."

Do you actually need a whole string of fossils to assess the relationships of living forms? For over a decade now there has been an increasing tendency among a certain school of systematists — to which I belong — not to view fossils as the be-all and end-all of determining evolutionary relationships but to incorporate them into such an analysis along with extant taxa as just additional sources of information. But Gregory's argument of 1922 still has quite a following. Many paleontologists and systematists believe that the real relationships among organisms can be determined only with fossils, and thus fossil hunting to fill in the "gaps" is of primary importance.

In addition to the obvious shortcoming of no fossil record, how, Gregory asked, would you "explain all the profound anatomical, physiological and psychological resemblances between existing Simiidae [apes] and Hominidae?" — resemblances not found between humans and tarsiers. You could, he suggested, conclude that all of these "profound resemblances" were due to convergent or parallel evolution in humans and apes. But obviously that was out of the question.

The evolutionary association of humans and tarsiers yielded messy scenarios that far exceeded the credulity of most systematists. The notion of there having been a brachiating background in human evolution — a gibbonlike ancestor — was a bit easier to accept, because the gibbon was obviously much closer evolutionarily to humans. Any animal could evolve to hold its torso erect, but there is something special about the human shoulder joint, also seen in the other hominoids, that would seem to attest to a brachiating past.

There is, however, another way to get to bipedal locomotion and erect posture: an animal on all fours could stand up. The great apes, a good Darwinian intermediate between four-footed monkeys and two-footed humans, seemed to be reasonable candidates for an evolutionary change of this kind.

When it was popular to think of the great apes as a real group of hominoid primates, there were some questions about the evolutionary story of human bipedalism that had to be explained. How and why did human bipedalism evolve from knuckle walking? And how and why did the orang-utan's "four-handed" arboreal double-jointedness evolve from the predominantly terrestrial form of knuckle walking typical of the chimpanzee and the gorilla? Vestiges of the African apes' type of knuckle walking were thought to remain in the cupped hand and foot of a grounded orang-utan, but it took just a few detailed studies to learn that only the African apes were true knuckle walkers, anatomically and functionally. Quite disturbingly, it was also quickly noticed that humans did not even have a veil of pseudo–knuckle walking. Somehow, all traces of our knuckle-walking heritage had been obliterated.

The argument for a knuckle-walking phase in the evolution of human bipedalism has reappeared with suggestions of the relatedness of humans to only the African apes. Darwin is cited, in the belief that the ensuing discussion will have greater validity. For, as Darwin had proposed, one can picture an evolutionary sequence going from the typical mammalian quadrupedal type of locomotion of an Old World monkey (for instance, a baboon or a rhesus macaque), to the semierect knuckle walking of an African ape, and finally to the fully erect, bipedal stance of a human. Thus, the knuckle walking of the African apes, with all of its anatomical attributes, must have characterized the last common ancestor of humans and the African apes.

This is the reason, the argument goes, we see these features in the living African apes: they inherited their knuckle walking from the ancestor they shared with humans. But why don't we see any evidence of this feature in the muscular or skeletal anatomy of humans? And if the specializations of knuckle walking are not unique to the African apes, because these specializations were inherited by them from a more distant ancestor, then what features *do* hold the African apes together evolutionarily?

There is a problem here. The curious thing is that the problem is rarely perceived.

Of course the African apes are sister taxa.

Why?

Because they are very similar animals.

Why?

Because they have anatomical specializations associated with knuckle walking.

Why?

Because they are sister taxa.

This circular reasoning illustrates how inconsistencies get lost in the rush to explain the evolutionary course of becoming human.

The "what ifs" and the "why fors" become even more entangled if you entertain the most recent (and rapidly becoming most favored) suggestion of hominoid relationships, in which the chimpanzee is thought to be more closely related to humans than to the gorilla. How do you explain knuckle walking and human bipedalism then?

One explanation that has been offered is that knuckle walking can still be taken as having united the larger group consisting of the gorilla, the chimpanzee, and humans. Gorillas and chimpanzees are functional and anatomical knuckle walkers because they inherited these attributes from this ultimate common ancestor. Humans lost these features. But then how did humans re-evolve the non-knuckle-walking muscular and skeletal anatomy typical of other primates?

Another scenario is that knuckle walking did not characterize anyone's ancestor. Knuckle walking evolved independently and in parallel — once in the gorilla and once in the chimpanzee. In this scenario there is no need to explain how and why humans lost all traces of a knuckle-walking ancestry. However, the idea that almost identical knuckle walking evolved independently in the chimpanzee and in the gorilla is difficult to defend in the face of the anatomy involved.

The special anatomy of the hand of the chimpanzee was known as early as 1862. Upon dissection of a chimpanzee, it was discovered that the tendons of the flexor muscles are surprisingly short. The flexor muscles arise on the palm side of the lower arm and send their tendons across the wrist and palm to attach on the undersides of the bones of the fingers. When you make a fist, pulling in or flexing your fingers, you can feel the contracting flexor muscles in your forearm. Flexors also pull in and bend your hand at the wrist joint. Extensor muscles on the outer side of the forearm straighten out or extend your hand and fingers. Humans and most other primates can extend their fingers at the same time they extend their hand by straightening out the wrist

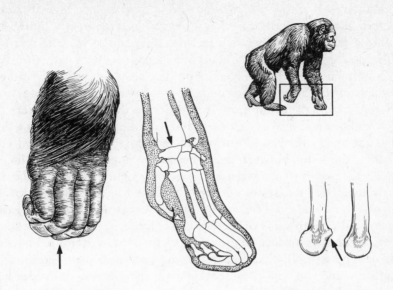

FIGURE 4B   A chimpanzee *(above)* in a typical knuckle-walking stance. Its hands are enlarged *(below)* to illustrate various anatomical specializations associated with knuckle walking in the chimpanzee and gorilla. Note on the fleshed-out hand the thickened friction pads on the weight-bearing surfaces of the fingers *(arrow)*. The fingers of the hand are also naturally flexed. On the bony hand, the lower, or distal, surface of the radius *(arrow)* is broad and the knuckle joints are bent inward, or hyperflexed. In this hyperflexion, the first phalanx of the finger rides farther over the articular surface of the metacarpal than it does, for example, in humans. In detail *(right)*, the articular surface of a metacarpal of a knuckle-walking chimpanzee or gorilla is enlarged to illustrate how it is modified *(arrow)* compared to the more common configuration, as seen in humans.

joint. Chimpanzees — and, as would later be found, gorillas — cannot. These two hominoids have such truncated flexors that the fingers cannot be straightened out when the hand is also extended. The short flexors will not allow this combined movement. Only when the hand is itself flexed can an African ape extend its fingers.

I hate to get into the "chicken or egg" disputes that are common in debates of things evolutionary, so I will refrain from speculating on whether knuckle walking evolved because the hands of an African ape were typically flexed to begin with or if the flexors became short in response to selection for knuckle walking. One can pursue questions

of relatedness without complicating matters with such probably un-
testable notions of how or why. Keep things simple. It is sufficient
to know what does or what does not exist, and, if it does, what it
looks like.

Russell Tuttle, of the University of Chicago, has for years studied
the anatomy of the hominoids with an eye toward understanding their
different modes of locomotion. In addition to the earlier discovery of
the foreshortened flexors, Tuttle found that the knuckle-walking African apes had other modifications of their forearms and hands. For
instance, the joints between the bones of the fingers and the palm as
well as the bones of the wrist are extremely tightly packed and held
firmly together by a system of well-developed ligaments. The lower
end of the radius, the bone of the forearm that articulates with the
bones on the thumb side of the wrist, is thickened, perhaps for added
strength, and its surface is grooved, so that the wrist of a knuckle
walker can essentially lock securely into position. If you tried for any
length of time to support your weight on the joints or knuckles between the longest finger bone and the shorter, second finger bone,
you would no doubt find it difficult because of the looseness in the
joints of your fingers, hands, and wrists.

Another aspect of the anatomy of knuckle walking is that the finger
bones that articulate with the metacarpals, the bones of the body of
the hand, bend backward at this joint when the knuckle-walking
posture is assumed. If you try to support yourself on your finger joints,
you will notice that the first finger bone (phalanx) will be aligned
with the metacarpal above it. If you were a chimpanzee or a gorilla,
the first phalangeal–metacarpal joint would be hyperextended and
bent backward. The upper end of the first phalanx would then articulate much farther around the lower end of the metacarpal. If we
were to look at this end of the metacarpal of a chimpanzee or a
gorilla, we would see that this area of greater articulation would be
distinctly enlarged.

If you try to support your weight on your phalangeal joints for
any length of time, you will probably begin to experience some pain.
Some of this is due to the fact that you just do not have the best
arrangement of tendons, ligaments, and bony contact surfaces. But
you also lack special pads of friction skin to cover and protect the
ends and the bodies of the phalanges that would be supporting your
weight and taking a beating from walking on the ground. Knuckle-
walking African apes have those special pads.

The anatomy of knuckle walking is something noteworthy. It seems to me that anyone claiming its presence in the gorilla and in the chimpanzee as anything other than a reflection of their phylogenetic closeness would need an even more strongly grounded alternative theory of relatedness — a theory grounded in some other very convincing biological evidence. But all too often, one theory of relatedness is ousted in favor of another not because there is a compelling biological reason, but because a more intriguing evolutionary argument has been proposed.

It has become very easy these days to dismiss a theory of relatedness by asserting either that the features that its proponent thinks suggest the unity of certain species are, in reality, merely features that came to resemble each other through independent and parallel courses, or that those features are just primitive retentions from some distant ancestor, whose distance from the present makes those features useless in the deciphering of relationships.

Claims of parallelism or primitive retention must be justified. Similarities shared by two organisms can be parallelisms only if other similarities, between one of these organisms and another organism, reflect phylogenetic closeness. Similarities between two organisms can be interpreted as primitive retentions only if other similarities, between one of the organisms and another organism, reflect more recent common ancestry. Unfortunately, these facts are often lost in the day-to-day struggle to build the better scenario.

In addition to disagreements over the best scenario for the origin of bipedalism, some of the most heated of human evolutionary debates have centered on the brain and three of its capacities: language, tool use, and tool manufacture. And the focus of many of these arguments was, of all things, the human canine tooth.

Modern humans, as well as fossil members of the genus *Homo*, have relatively small canines, which do not project very far beyond the cutting edges of the neighboring teeth. In contrast, baboons, for example, especially the males, have huge, daggerlike canines. The upper canines in particular can present quite a threatening picture when a male baboon just yawns in your direction. The large apes, although they have less fierce-looking canine teeth, do have canines much stouter and taller than our own. Large canines serve two functions. One is defense and attack, real or simply threatened. And the other is feeding, whether it is in hunting and killing prey or in tearing and processing food.

Ever since Darwin, evolutionists have been painfully aware of just how physically puny and defenseless the human is. Why would natural selection proceed along an evolutionary course that would yield such a helpless creature? Not only are humans physically weak, they do not even have a set of large canines with which to threaten a potential enemy.

The explanation appears simple: early hominids developed the ability to use and then to make tools. And tools can take the place of large canines for defense as well as for the procurement and preparation of food.

But as it stands, the explanation is incomplete, for humans are not the only animals that use tools. The otter carries its shell-cracking stone with it everywhere, ready for use upon discovery of a cache of mollusks. Even crows, immortalized in Aesop's fables for dropping pebbles into a small well of water to force the water up to the top so it could be drunk, use tools. The three large apes have been observed using objects about them — twigs, branches, leaves — as tools, but there is unavoidable evidence that these primates are distinct from the average tool-using animal in that they also modify objects to be used as tools. These apes, then, have crossed a threshold that was thought to be a boundary of humanness, and that is the ability to create a functional object. But none of them has gone so far as we have. The apes still have large canines.

The more complete explanation of our canine reduction is based on the idea that the evolution of bipedalism freed our hands for tool business. With tools for hunting and food preparation as well as defense, humans did not need large canines, which set into motion selection for canine reduction. If you are going to replace teeth with tools, then the art of tool making had better proceed apace with canine size reduction. The biofeedback system thus begun is, ultimately, tied to the brain. With the freeing of the hands from a role in locomotion to the task of using and manufacturing tools, there was an increased stimulus for brain development, which in turn allowed for the creation of more sophisticated tools and the appearance of other uniquely human behaviors such as language.

This scenario of canine size reduction in hominid evolution had an impact upon how one interpreted the potential early hominid ancestors. How much of a ready-made human ancestor could one expect to find?

There was *Ramapithecus,* with its rather flat molar chewing surfaces and its somewhat small canine — at least, *Ramapithecus'* upper canine looked small compared with that of an ape. But since humans were supposed to have evolved from the apes, or something apelike, anyway, *Ramapithecus* could fit quite well into a graded series that proceeded through the gracile species of *Australopithecus,* with its canine representing an intermediate size, and eventually resulted in the genus *Homo* and its most reduced canine.

On the other hand, there was *Gigantopithecus.* This fossil hominoid seems to have had a very small canine, which, in addition to being more like ours than *Ramapithecus'* canine in relative size, appears to have been incorporated into a functional complex with the anterior teeth of the jaw, the incisors. This is definitely *Homo*-like. However, if *Gigantopithecus* was the earliest hominid, then we would have to trace the ancestry of the genus *Homo* to the robust, and especially the hyper-robust, type of *Australopithecus,* because this hominid had a tiny canine that was at least spatially, if not functionally, aligned with the incisors.

What a mess. The idea that a massive, beetle-browed, big-jawed brute could be our ancestor was almost too much to bear. How could something like Louis Leakey's Zinjanthropus give rise to the genus *Homo?* Obviously, the possession by *Gigantopithecus* and the robust types of *Australopithecus* of *Homo*-like small and functionally incisorlike canines could not be phylogenetically significant. These attributes must have been acquired independently and in parallel by these different hominoids, since they were not present in the gracile *Australopithecus,* which, because of its thinner bones, more rounded head, and smaller cheek teeth, was assumed to be the best candidate for the ancestor of the genus *Homo.* We know, after all, that thinner cranial bones and more rounded skulls are somehow tied into the evolution of the human brain.

So the battle of the scenarios on canine size reduction, the origin of bipedalism, and brain evolution — basically, why some hominoids were in our ancestry and others were not — eventually died down. Then Johanson and his colleagues found Lucy, with its large canines, small but thickened cranium, and massive molars. Once again, these issues became the center of heated encounters.

Debates over which fossil is the real ancestor may be popular, but they can be as ephemeral as the winds that expose the fossil that will next receive such attention. The most important endeavor — but the one most frequently brushed over — is how, in the first place, any scheme of relatedness is determined.

We all have the urge to find order in the natural world. We try to make sense of the world around us by looking for ways in which to group small units into larger, identifiable units, and these, in turn, into larger and larger, but also meaningful, assemblages.

We try to minimize the "gray zones" between things, to diminish the unknown. This propensity was pointed to by the British ethno-linguist Sir Edmund Leach in a provocative article on pejorative language and animal abuse in language. Leach argued that the animal names we use when swearing (for example, bullshit) are typically the names of domesticated animals, and domesticated animals can be thought of as occupying a disturbing, unresolved "gray zone" in our minds. These animals are neither members of the family, as is a pet, nor distinctly removed from us, as is a wild animal.

The predisposition to group organisms in a meaningful way is certainly not limited to the most recent generations of human beings. In fact, what might be called early taxonomies, or classifications, of animals are known from ancient Assyrian tablets. Perhaps the most well known early attempt at organizing the organic world is the Laws of Kosher or Cleanliness, presented in Leviticus 11 in the Old Testament:

> And the LORD said to Moses and Aaron, "Say to the people of Israel, These are the living things which you may eat among all the beasts that are on the earth. Whatever parts the hoof and is cloven-footed and chews the cud, among the animals, you may eat. Nevertheless among those that chew the cud or part the hoof, you shall not eat these: The camel, because it chews the cud but does not part the hoof, is unclean to you. And the rock badger, because it chews the cud but does not part the hoof, is unclean to you. And the hare, because it chews the cud but does not part the hoof, is unclean to you. Of their flesh you shall not eat, and their carcasses you shall not touch; they are unclean to you."

The passage continues with references to a plethora of organisms. Fish, for example, are divided into those with scales and those without, such as would characterize, for example, a tuna versus an eel. Some birds are considered clean and thus edible, but others, including the eagle, the osprey, all types of ravens, the ostrich, the owl, the carrion vulture, and the bat (yes, the bat), are not. In addition to these animals, "every swarming thing that swarms upon the earth is an abomination" and thus, it probably goes without saying, "shall not be eaten." A swarming animal is "whatever goes on its belly, and whatever goes on all fours, or whatever has many feet."

With our historical advantage, we can isolate a few errors in these statements in Leviticus. For instance, the camel is actually similar to "clean" cloven-footed animals, such as the cow and sheep, in the development of its feet. Cloven-footed animals have only two bones between the wrist bones and the upper bones of the forefoot, and between the ankle bones and the upper bones of the hindfoot. Each of these bones, called metacarpals in the forefoot and metatarsals in the hindfoot, corresponds to one toe. All primates have five metacarpals in each hand and five metatarsals in each foot. A cow or a camel has only two toes per foot, and thus has only two metacarpals or two metatarsals per foot.

In the camel — and in the cow and the sheep — each pair of metacarpals and each pair of metatarsals fuses along much of its length, creating what is essentially one bone, the so-called cannon bone, that remains "split" at its lower end. It is this split that gives the animal the cloven hoof. The camel is certainly a cloven-footed animal. The major difference in the foot between a camel and a cow or sheep is that a camel's toes bear soft pads rather than an expanded, hooflike nail.

Although they share cloven-footedness with cows and sheep, camels are not as fully anatomically equipped to chew the cud as these true ruminants. Chewing the cud, or ruminating, involves rechewing food as it makes a regurgitative journey through each of four digestive chambers of the stomach. The camel, however, has a three-chambered stomach.

If the anatomy of camels had been properly understood, would it have made a difference in the acceptability of the camel as food? Probably not. Given its value as a beast of burden, I think some taboo would have been created to protect the camel from being consumed, at least prematurely. (I say "at least prematurely" because during the analysis of animal bones from fifth-to-eighth-century Vandal/Byzantine deposits of the famous Tunisian city of Carthage, I found that after horses, mules, and camels had been worked hard enough and long enough to cripple them with arthritis, they were butchered and probably eaten.)

These confusions about the camel's place in nature stemmed from mistaken notions of its anatomy. Considering how many centuries ago Leviticus was composed, however, the attempt to understand detail is impressive. For example, the "rock badger," which must be what today we call the rock hyrax of the Levantine region, constantly works its jaws as if chewing its cud. This strange little animal, which does look somewhat like a badger but with fangs, has puzzled systematists for decades. In fact, the ridging on its molars, and especially its funny little feet with their "hoof-capped" toes, have prompted speculation that this rabbit-sized animal may be most closely related to the elephant.

Obviously, even if we find cause to disagree about the details, at least it is clear that morphological features were used at that time to specify groups of animals. But other criteria were also brought to bear — in Leviticus and other old writings — on the grouping of organisms. One of these nonmorphological attributes was habitus — that is, where one finds an organism. For centuries, whales were grouped with fish because they are aquatic animals. Bats were considered to be birds because they travel by air. Along with habitus, and at times almost synonymous with it, is behavior. "Swarming" had its place in Leviticus. And group dynamics have indeed played a central role in how some more recent taxonomists have grouped animals. Dietary proclivities, such as seeking out specific species of insects or the pollen of certain plants, have also come to have a place in classificatory schemes.

Although these three criteria — morphology, habitus, and behavior — persist in their influence upon taxonomic pursuits, most modern taxonomists would agree that the most profitable avenue of investigation is the study of morphology. Habitus is merely the environment in which an organism finds that it can get around. The "fins" of a whale are not, in their structure, the same as the fins of a shark or a bass. The "wing" of a bat, in addition to lacking a coat of feathers, is not constructed of bones in exactly the same configuration as in the wing of a bird. Even more open to criticism as a criterion by which to group organisms is behavior. Animals that are physically very different can have strikingly similar behavioral repertoires.

Agreeing that it might be worthwhile to study the morphology of organisms to determine natural groups is, however, easier than doing it. All too often systematists lose sight of what they are trying to accomplish. The sheer effort of lumping individuals into groups, and then creating groups of groups, seems to cloud the purpose of such activity. The goal has often become the generation of a classification of the organisms under study.

But any classification is just names, lists of names. A group of individuals is called a species and is given its own name. Groups of groups are at higher levels in the classification, and they all get named. A genus subsumes one or more species, a family contains one or more genera, families may make up a so-called superfamily or infraorder, infraorders or suborders compose orders, and so on. You, the classifier, ultimately determine not just the name but the ranks of each group and the number of different groups (taxa) that you will allow into each rank. It is often forgotten, when one is engaged in debates on what is a primate or a hominid or an anything else, that classifications are typically the amalgamation of many types of information. Some of what goes into a classification is truly objective or at least testable, but much of it is quite subjective and even peculiar to a particular taxonomist.

Classifications quite frequently convey all sorts of meanings other than the apparent natural grouping of organisms. Consider, for example, the family of "man," Hominidae. Who are the hominids? That depends on who you happen to be reading, or believing, at the moment.

When Raymond Dart concluded his 1925 *Nature* paper on *Australopithecus africanus,* he proposed a new taxonomic family to contain the new human ancestor: Homo-simiadae. He thought that his new ancestor, the intermediate form between the apes and the genus *Homo,* was deserving of the taxonomic recognition that isolation in a separate family would afford. Thus, *Australopithecus* would be emphasized as something special, or at least something different, by being placed in a taxonomic rank of some stature. This taxonomic move may also have been prompted by a reluctance to be even more bold and outrageous than he had already been with his assertions about his new genus, *Australopithecus*. Putting *Australopithecus* into a separate family could give the appearance of separating the genus from its presumed ancestor, *Homo*.

With the discovery of more and different types of *Australopithecus* and the more widespread recognition of the species of this genus as early representatives of our own lineage, the family Hominidae came to be occupied by *Australopithecus* and *Homo*. After all, one species of *Australopithecus* must have been *the* ancestor of the first *Homo*.

However, so as not to create the illusion of too much closeness, or of deep, cerebral similarity between an "ape-man" and a "true man," the single genus *Australopithecus* was placed in a separate subfamily, Australopithecinae. The lone genus *Homo,* with its few potential species, was protected in its own subfamily, Homininae. Separateness was still maintained, but at least *Australopithecus* was brought a little closer to its apparent descendant.

With Elwyn Simons's resurrection of *Ramapithecus* as a viable candidate for an even earlier ancestor in the human lineage, the family called Hominidae had to be rethought. What does one do with an ancestor? This problem has in general received quite a bit of attention, especially from the American paleontologist who dominated the field for over fifty years, George Gaylord Simpson.

The problem of how to deal with ancestors in a classification exists as a result of actually having tried to identify some organisms as the ancestors of others. Since an ancestor is, in effect, supposed to be a transitional form in a continuum, it would seem to be arbitrary where one draws the classificatory line. Do you include the ancestor in the same group with its descendants? Should you keep an ancestor with the group from which it evolved? Or should you, as Dart did with *Australopithecus,* put the ancestor into a classificatory group, or taxon, of its own?

Ramapithecus was Simons's ancestor of the proper hominids, and he included *Ramapithecus* in the family Hominidae along with *Australopithecus* and *Homo.* But to keep some semblance of separateness, the subfamily Ramapithecinae was sometimes used to refer to the single genus it housed. In the 1970s, when it became apparent that *Sivapithecus* and *Gigantopithecus* should be dissociated from *Dryopithecus* and aligned with *Ramapithecus,* David Pilbeam proposed that these three dentally hominid hominoids should be relegated to their own family, Ramapithecidae, even though these three taxa were also thought to be intimately related to subsequent hominids. The family Ramapithecidae essentially emphasized the distinctiveness of the earlier from the later hominids.

There is something almost magical about classifications. They are powerful statements. But at the same time they can be so vague that they are virtually undecipherable. They are really most often expressions of the classifier's world views — who should sit where, classificatorily, relative to the classifier's genus and species. The debate over whether a fossil is or is not a true member of *Homo sapiens* tends to sound more like a debate on the admission of an underclassmate to a fraternity than it does a rigorous discourse on the specimen's phylogenetic affinities.

One of the best examples of the arbitrariness of classificatory schemes is the use and abuse of the family Hominidae. Contrast the constituents of the family Hominidae with, for instance, those of the family Cercopithecidae, the Old World monkeys. This family includes the living baboons, mandrills, macaques, patas monkeys, guerezas, and langurs, as well as the many genera of fossil baboons and other monkeys, some of whom were contemporaries of *Australopithecus*. Without even considering the number of species involved, there are almost *thirty* genera of fossil and extant cercopithecids. Thus, depending on which classification of Hominidae one chooses to follow, there are anywhere from over five to almost ten times as many Old World monkey genera crammed into a single taxonomic family as there are hominid genera. And the curious thing is that the diversity among Old World monkeys, whether or not you care to talk about the now extinct giants, is certainly much greater than you would see if you put all potential hominids into the same family.

Why does such lopsidedness exist? The usually unspoken reason is that there is something "special" about being human and about being a human ancestor. And this something special is emphasized by the classificatory elevation of this handful of taxa to the rank of family. It might be interesting to be an Old World monkey, but, obviously, being a baboon or a langur is not as evolutionarily significant as being a human or a near-human primate.

The use of taxonomic rank either to elevate or to diminish an organism is done quite frequently to make a point. However, in most cases, the basis of that point is known only to the classifier. Consider, for example, the aye-aye of Madagascar.

The aye-aye, whose genus is *Daubentonia,* is a very strange-looking nocturnal primate that has bedeviled systematists for well over a century. To begin with, the aye-aye is rather homely, with oversized ears, long, scraggly hair, and very peculiar and elongate fingers, of which the middle digit is thin and scrawny, tipped with large, curved claws. On the ground, the aye-aye looks almost pathetic as it tries to walk without tripping over its hands.

In its teeth, the aye-aye is also peculiar among primates. Instead

FIGURE 4C   The aye-aye of Madagascar, *Daubentonia madagascariensis*. While primate in having, for example, a flattened nail on its big toe, the aye-aye has been a most troublesome primate to deal with systematically, not only because of its unusual appearance, with its elongate fingers and toes tipped with claws, its scraggly hair and oversized ears, but because, like a rodent, it has a large pair of anterior teeth that continue to grow and replace themselves throughout the animal's life.

of having a full set of teeth distributed along its jaws as, for instance, you and I have, the aye-aye develops only a pair of teeth at the front of the upper and the lower jaws, behind which is a toothless gap, and then a set of puny cheek teeth with essentially featureless chewing surfaces. To make things even more puzzling, the upper and lower pairs of teeth are large and continue to grow throughout the lifetime of the animal. In its jaws and teeth, *Daubentonia* is most like a large, gnawing rodent — a beaver, for example, which also has anterior teeth that continually grow and replenish themselves as they are worn down through chewing tough, woody plants.

Daubentonia has been interpreted in two diametrically opposite ways. One interpretation is that it must be an aberrant evolutionary holdover. This notion gained some support from comparison with the most ancient of fossil primates (known from Paleocene deposits in Europe and North America perhaps as old as sixty-five million years), which are characterized by having an enlarged pair of anterior teeth. One of these Paleocene primates was most like the aye-aye in having an extremely deep and stout mandible and an equally stout anterior tooth. That fossil primate was given the genus name *Chiromyoides* after the genus then used for the aye-aye, *Chiromys*.

A result of perceiving the aye-aye as an aberrant evolutionary holdover was to place it not only in its own family, Daubentoniidae, but also in its own superfamily, Daubentonioidea. So for decades the lone genus and species of the living aye-aye, *Daubentonia madagascariensis,* lay in taxonomic limbo, the victim of hyper-classificatory vigor. All one could derive from looking at a classification of primates was that *Daubentonia* was separated from all other "lower" primates; one never knew why. A further, and perhaps the most important, consequence of this taxonomic exile was that the aye-aye was so effectively isolated from other genera by its high rank that one never quite understood the animal's possible relationships within the broader group of primates.

In the second interpretation, instead of perceiving *Daubentonia* as an evolutionary holdover, some systematists argued that this small, odd animal is really quite specialized. It is, after all, the only rodentlike primate. Those comparisons of the aye-aye to Paleocene fossil primates are just superficial, the second group of systematists pointed out. Radiographs of the fossils' teeth and jaws reveal that the root tips of the enlarged anterior teeth are closed off, which means that their teeth, like ours, stopped growing once they had erupted into the jaw. However, the root tips of the anterior teeth of the aye-aye remain open and actively produce more tooth throughout the life of the animal. Because of its specializations, the aye-aye was put into its own family, Daubentoniidae, and even into its own superfamily, Daubentonioidea.

Whether one sees the aye-aye as an evolutionary holdover, linked to the Paleocene fossil primates, or as an organism unique among living primates, the effect is the same: the primate has been separated taxonomically from the members of its larger evolutionary group. Ei-

ther way, you would not know what the aye-aye's phylogenetic relationships were.

Classifications should be for communication. How one goes about
generating one, and what one wants to communicate in it, are different
matters altogether. But the reason that a classification typically holds
such sway over us is that we are inclined to imbue it with the authority
of a phylogeny. That is, we assume it has evolutionary significance.
The situation gets even more complicated when we try to fit in all
possible ancestors and use them to examine the whys and wherefores
of an animal's evolutionary past. Unfortunately, the identity of ancestors and events of the distant past are most often illusory, and their
purported details are highly susceptible to the subjectivity of the systematist.

For example, the ancestry of the genus *Homo* is still up for grabs.
Louis Leakey and subsequently his son Richard would seek a long
separation of *Homo* and *Australopithecus,* implying that the two
groups had distinct evolutionary pasts after departing from a common
ancestor. Don Johanson and his colleagues see in their Lucy, *Australopithecus afarensis,* the ancestor of all other *Australopithecus* as
well as the genus *Homo.* Some paleoanthropologists, however, do
not think that *A. afarensis* is different from other *Australopithecus*
and thus believe that the species *afarensis* should be sunk. The pre-
dominant view has been that *A. africanus* is the ancestor of other
Australopithecus as well as *Homo.* But there was a brief time when
the hyper-robust type of *Australopithecus, A. boisei,* was thought to
be the ancestor. Nevertheless, and in the face of a history of deposed
candidates for human ancestry, the pursuit of human ancestors continues to preoccupy most paleoanthropologists, whether they are the
discoverers or the interpreters. In school we all learn the paleontological dictum "The fossil record is only as good as your last field season" — and then, unfortunately, we usually forget it.

I am not suggesting that ancestors did not exist. However, it is
difficult, at best, to know when you have found one.

How "good" an ancestor could yours be, anyway, if I can go out
and, with luck, find an older fossil, which I then declare to be *the*
ancestor? If you wanted to get the throne of ancestry back into your
lab, would you have to find an even older specimen than mine? What
if you or I actually found something that looked more the way we
would expect the ancestor to look, but it was in more recent deposits?

The identification of ancestors is based primarily on the notion that ancientness equates with primitiveness. We expect an ancestor to be more primitive than its descendant. Thus, if we agree that we are dealing with hominids, the best candidates for ultimate ancestry are those that come from the most ancient deposits. Because they are so old, they must be more primitive in their features than specimens found in younger deposits, which in turn must be more derived — that is, more specialized, more developed evolutionarily.

With this, albeit circular, justification, much of the effort of unraveling the relationships of organisms is basically a matter of discovery. We must find more fossils, of increasing age, to fill in the gaps, so that we can observe the details of evolution and deepen our temporal perspective.

What if there is more than one fossil in a stratum? How does one decide which is the ancestor and which is the side branch? This is usually resolved by assuming that the fossil that looks the most similar overall to forms found in younger strata must be their ancestor.

The two operational criteria that have most commonly been used to reconstruct the evolutionary relationships among organisms are overall similarity and time. Time performs two functions. First, it places a given species into a larger picture, where the chronology of taxa can then be observed. Second, it creates a framework within which organisms can be compared. If your specimen is old, then you know that its features must represent less advanced evolutionary states than you see in younger specimens. One of the most difficult problems facing systematists is not knowing the age of a particular specimen. Without knowing how old a taxon is, and thus where it sits relative to other taxa, many systematists feel that you cannot make any reliable judgment about its ancestry and descent and the primitiveness or derivedness of its morphologies. If there are two or more competing ancestors, the one that looks more like the younger taxa is at least in the main line of evolution.

As Elwyn Simons once told me, most paleontologists are essentially historians trying to discover evolutionary relationships by studying the past. And there does seem to be something intuitively correct about believing that a historical approach — finding ancestors and missing links — will eventually prove worthwhile. But what does one do in the meantime, until those last missing links are discovered?

Analyzing fossils is a difficult business — but perhaps not as difficult as becoming a fossil. First, you would probably have to die in or near water, preferably in slow-moving water like a lake bottom so that your bones don't get battered around and can become hardened with minerals. You have to be safe from forces that would erode or shatter your fossilized bits and pieces into smaller bits and pieces, but you don't want to be completely covered up or no one would ever find you. When the time of discovery approaches, you need to be exposed enough to be spied, but not for such a long time that wind-blown sands or changes in temperature would turn you into a puff of dust. Given all the difficulties, is it any wonder that the paleontologists may never piece together the whole picture?

It is the end of a grueling field season — almost no fossils, and definitely nothing new and different, to show for all your labors. Does that mean you cannot continue to work on resolving the relationships of your pet organisms? What if it turned out that your favorite animals had no retrievable fossil record — not that a fossil record could not have existed, but that, as is the case with birds or insects, problems of preservation and sampling intruded upon the paleontological recovery of specimens? Do you just give up and go home? Of course not.

Even without fossils, we know that all living things are related, somehow, to an ultimate, last common ancestor they all shared at the beginning. And we know that some living organisms are more closely related to each other than to others. Every living organism has its closest living relative, and these sisters have their closest living relative, and so on.

Let us say that you want to figure out the relationships among three insects, moths, perhaps. Because they are insects, they have a very poor fossil record. In fact, predecessors of these particular moths have so far eluded discovery, either in amber, the fossilized resin of extinct trees, or in shales, the thin layering of sediments that quietly settle to the bottom of calm waters and bury the bodies of the dead.

But as an ambitious systematist, you remain undaunted and scrutinize in great detail the antennae, wings, hairs, body segments, and so on of these moths. After weeks of such tedious work, you tally up your data and discover that two of the moths are particularly similar to each other in certain aspects of their antennae and wings. At this

FIGURE 4 D Three hypothetical moths, each representing, for example, a species. Although all three moths are patterned differently, two of them are otherwise quite similar overall — for example, in the shapes of their wings and antennae — while the third is dissimilar in wing and antenna configuration. Because the moths on the left and in the middle are the most similar, many systematists would conclude that they are also closely related. But there is not enough information here to make that judgment.

point, it might seem that you have done your job. You triumphantly conclude that you have solved the case. Obviously, you say, the two most similar moths are the more closely related of the three.

However, you have really only begun your analysis. In order for you to demonstrate the closer evolutionary relationship of two of the three taxa you are immediately interested in, you have to convince me that the similarities you have drawn my attention to are indeed features unique to the organisms that share them.

One slant on the problem is to look at things from the perspective of the third moth — the odd taxon out. If the two moths or taxa you think are closely related really are, the one excluded from further relatedness within its group must have remained primitive, or unchanged, in the general structure of the organs or features you are looking at. From the standpoint of the third moth or taxon, the other two must appear derived, different, and changed in details of the same organs or features. The two derived taxa are then plausibly united by virtue of their ancestor having changed and become different. Their apparent evolutionary novelties attest to their relatedness.

Thinking to yourself that that is precisely what you demonstrated, you are probably wondering what I am jabbering about. Two taxa were more similar to each other than either was to the third in the study. Obviously, the third one was different because it remained primitive — it did not change. The other two *did* change — or at least their common ancestor did — and that is why they don't look like the third one.

But the only way that you can demonstrate this, and demonstrate precisely what the significance of difference or similarity is, is within the context of a broad comparison, a comparison that must be larger than the few taxa you might be most interested in. That is, you must demonstrate that the reason the third taxon is primitive (relative to the two you find the most similar and thus think are the most closely related) is that the third taxon is basically the same as many other organisms, and thus it has to have remained relatively unchanged. If this is the case, it will certainly become apparent if this organism does not look very different from a slew of other organisms.

In the context of the broader comparison, the two taxa that look most similar to each other might truly emerge as unique in their similarities. If they do stand out so vividly in shared novelties, there will be a stronger case for suggesting their close evolutionary relationship.

FIGURE 4E   In a broader comparative context, the two moths that in figure 4D are similar get lost in the crowd of other moths that are equally similar in the shapes of their wings and antennae. However, the third moth in figure 4D stands out in the crowd of moths because of the unique features of its wings and antennae. Overall similarity does not necessarily indicate a close relationship. One must look for those uniquenesses that make a species or a group of species stand out from the masses.

But what if the broader comparison demonstrates that the third taxon does not blend in with the background morphology of the masses? What if, instead, the two taxa that appear most similar to each other get swallowed up in the broader comparison? That these two organisms turn out to be pretty much like everything else? And that, to your astonishment, that supposedly primitive third taxon almost leaps out at you because of its relative distinctiveness, its uniquenesses, its novelties? The broader comparison reveals that similarity between two of your original three taxa does not demonstrate evolutionary closeness between the two because they are also similar to an array of other taxa. Instead of relatedness, the broader comparison in this instance points to the distinctiveness of the third original taxon. This often happens if you reject the initial impulse to draw conclusions of relatedness based on study of only a few taxa and take the time to tease apart the nature of overall similarity. Only by making comparisons among an array of taxa can you delineate those features specific to your taxa of primary interest.

Uniqueness is the clue. Uniqueness is what makes a species distinct, even if it is just by a hair on an antenna. And uniqueness distinguishes the closely related, because their distinctions are the uniquenesses of their last common ancestor — an ancestor that only they share. Given a large group of related organisms (mammals, perhaps), each successive ancestor of each subsequently smaller group of mammals will be distinguished by at least one feature unique to it. It is a systematist's obligation to try to weed out, from the totality of features that make up an organism, those features or that feature unique to it and its closest relatives.

This approach to elucidating evolutionary relationships, commonly called cladism, has been around in sentiment and theory for quite a while. In practice, however, many systematists fall back on the more comfortable criteria of overall similarity and, when available, the chronology of the fossil record.

Whether the organisms you are studying are living forms or fossils, you can make the same statement about their possible phylogenetic relationships. If they are truly related as members of a larger evolutionary group, or clade, then two of them will be more closely related to each other than either will be to any third organism.

But, you might ask, what if one of the fossil forms was actually the ancestor of one or more of the living ones? Aside from daring you to prove that any fossil was actually the ancestor of anything else, I would respond by saying that, if this was the case, the simple statement of relatedness my cladistic approach yields would not violate reality.

If a fossil form really was the ancestor of another known organism and you were fortunate enough to have the necessary comparative information to generate a theory of relatedness that grouped these two as sister taxa, you would still be on safe evolutionary ground. If you really are working with an ancestor and its descendant, then declaring that the two are more closely related to each other than to anything else is a true statement. And if you did convincingly demonstrate first that a fossil and a living form were sister taxa, you might be more likely to argue well for an ancestor–descendant relationship between the two than if you simply jumped in and bestowed ancestry upon a taxon because it was the oldest thing you had on hand.

A theory of relatedness that is based first and foremost on biological aspects of organisms, and which generates the simplest statements of relatedness, is the easiest to corroborate or falsify, or, if you prefer, to prove or disprove. Scenarios, like arguments for ancestry and descent, are farther away from the data, and, while they may perhaps be more fun to generate than other evolutionary formulations, they are less susceptible to scrutiny and possible rejection.

It makes some sense that one should not invest too heavily in the resolving power of the fossil record in matters of relatedness. The fossil record is indeed only as good as the last field season, and there is nothing to ensure that you have sampled enough, or that enough of what existed even got preserved in the first place.

Another stumbling block in evolutionary studies is dealing with the concept of difference. What is the significance of the difference between organisms? A good example of the problem is the way in which the aye-aye was interpreted. This strange-looking animal was thought of as primitive in large part because it is so different from all other living primates. However, in its continually growing anterior teeth and its incredibly elongate and spindly fingers, the aye-aye is actually among the most unusual of primates and is best interpreted as being one of the most evolutionarily derived, or specialized, of primates.

The example of the aye-aye is not meant to imply that difference between organisms never reflects primitiveness. The question is, however, who is primitive relative to whom? In various features, the aye-aye is unique relative to all other primates. It is really these primates that are primitive in these various features relative to the aye-aye. But all too often we get caught up in trying to figure out relationships by merely looking at how different everything else is from our favorite animal.

Consider, for example, the reasons for thinking that one ape rather than another was our closest evolutionary relative. As Huxley pointed out over a century ago, the gorilla is the most similar overall to humans; then comes the chimpanzee, and then the orang-utan. G. Elliot Smith took overall similarity as a direct reflection of evolutionary closeness and concluded that the gorilla was the most closely related to humans. Although William K. Gregory pointed to the same overall similarity between humans and the gorilla, he concluded that humans were actually closely related to both African apes because the African apes themselves were intimately related (judging from their overwhelming degree of similarity). However, there are three competing theories of relatedness here.

One theory is that the chimpanzee and the gorilla are sister taxa by virtue of their many similarities. However, if this is the case, then those similarities between humans and gorillas and between humans and chimpanzees must have arisen independently and in parallel in humans and gorillas and in humans and chimpanzees. The second scheme is that humans and gorillas are related. But if this is correct, then features shared by gorillas and chimpanzees, as well as features shared by humans and chimpanzees, must have evolved independently and in parallel. The same interpretation would apply to the features shared by humans and gorillas as well as by gorillas and chimpanzees if, in the third possible case, the features shared by humans and chimpanzees indeed reflect their relatedness.

A major consequence of Gregory's endeavors was that, of the three great apes, the orang-utan was perceived as being less closely related to humans than is either of the African apes. But this interpretation was based primarily on the observation that the orang-utan shares fewer features overall with humans than does either of the African apes.

If one were merely to count numbers of features, the orang-utan would emerge as sharing fewer, overall, with humans than would a chimpanzee or a gorilla. But the reason the orang-utan emerges as more dissimilar to humans than either of the African apes is because it is unique among hominoids in *not having* many of their typical morphologies.

For example, some orang-utans do not have a nail on their big toe or thumb. In other orang-utans, these digits may also be quite rudimentary and underdeveloped. Thus, if one were to construct a list of similarities between humans and each of the great apes, humans would end up sharing with the gorilla: (1) nail on thumb, (2) nail on big toe, (3) well-developed thumb, (4) well-developed big toe. The list of human–chimpanzee similarities would be modified slightly to reflect the fact that the thumb, especially, is more reduced and situated farther away from the other digits of the hand. Depending on which variety of orang-utan was chosen for study, the number of human–orang-utan similarities could be reduced to zero.

Among the large hominoids, the orang-utan, in aspects of its hand and foot, is the most different from, and thus the least similar to, humans. But this does not mean that the orang-utan is the most

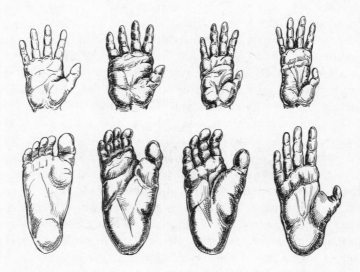

FIGURE 4F Hands *(above)* and feet *(below)* of *(from left to right)* a human, a gorilla, a chimpanzee, and an orang-utan (all drawn to the same length). On the basis solely of comparison of overall similarity, the human and the gorilla appear most alike and the human and the orang-utan least alike. But it need not be the case that overall similarity and close relatedness are equivalent.

primitive of the apes and therefore less closely related to humans. Compared to all hominoids, as well as to virtually all other primates, the orang-utan emerges as *unique* in these features of hand and foot. The difference, then, is not caused by the orang-utan's being more primitive, less evolutionarily far along than the other hominoids. Rather, and in contrast to the other hominoids — humans, gorillas, chimpanzees, and gibbons — the orang-utan is more derived, more specialized, in aspects of its hand and foot, such as reduction of the thumb and big toe and lack of a nail on the big toe and sometimes also on the thumb.

The double-jointedness of the hip of the orang-utan is due to the fact that a ligament, the teres ligament, which holds the femur firmly in the hip socket in other primates, is lacking. If we added the presence or absence of this ligament to our list of comparison between humans and other hominoids, the orang-utan would again emerge as less like humans than the gorilla or the chimpanzee. But, as with the comparison of the orang-utan's thumb and big toe, the orang-utan would also be less similar to humans than a gibbon or a baboon would be. The broader comparison reveals once more that the orang-utan is unique, not primitive.

The orang-utan has been penalized not only because it deviated from the typical hominoid condition by reducing or losing structures, but because it became different through the addition or embellishment of features. For example, orang-utans, especially the males, develop large throat sacs. Males also develop large cheek flanges. None of these features characterizes other male primates, much less other male hominoids. Orang-utans also grow the longest body hair among the primates, and the enamel on an orang-utan's cheek teeth is the most heavily wrinkled.

As our list grows, so does the impression that the orang-utan, because it is less similar to humans than are the African apes, is less closely related to humans than is either of the African apes. But that is certainly not the only interpretation of the data.

How one goes about making comparisons among organisms will greatly affect the resultant theory of relationships. For example, a common procedure is to arrange, in descending order, and by degree of similarity, organisms that have been compared with the focal organism of the study. Thus, if humans are the focus, the orang-utan would be farther removed than the African apes. But let's ask the

FIGURE 4G   *(Above)* Perception of how closely primates are related to humans, based on overall similarities, from the perspective of a human, such as Thomas Huxley or Adolph Schultz. In this world view, humans may be closest to the great apes, but humans are still a bit distanced from the apes. The great apes emerge as a group, with the two African apes closer to each other than to the orang-utan but also seemingly more similar to humans. Of the two African apes, the gorilla is slightly closer to humans. In the background is the gibbon, then Old and New World monkeys, the tarsier, and a lemur.

(Opposite) Perception of how closely primates are related to the orang-utan, based on overall similarity, from the perspective of an orang-utan. The other great apes appear quite close, with the chimpanzee actually being closer than the gorilla. This great ape group is set apart from the other primates, including humans. The gibbon, then the monkeys, and prosimians such as the tarsier, lemur, and bushbaby trail off into the background.

question differently. What would the result be if one took the orang-utan as the focal organism?

If we arrange our organisms relative to the orang-utan, the other hominoids would be strung out in order of increasing dissimilarity with the result that humans would be farther away from the red ape than would be the African apes. But the African apes would be closer to the red ape than they would be to humans. For example, the three large apes are similar in having relatively elongate arms and relatively short legs, whereas the reverse in limb proportions characterizes hu-

mans. Indeed, we would find, as did the Swiss primatologist Adolph Schultz, that although humans might appear more similar overall to the gorilla among the hominoids, the great apes themselves are even more similar overall to each other. This was the basis for Schultz's claim of having demonstrated, beyond a doubt, the phylogenetic unity of the great apes.

The essence of relatedness is that organisms which are closely related will be similar in features that they do not also share with less closely related organisms. These are the similarities of immediate heritage, of recent common ancestry.

Perhaps, if it is not first driven to extinction, the aye-aye will become the ancestor of a larger group of aye-aye-like primates. The descendants will inherit from their aye-aye ancestor at least some of the characteristics that made the aye-aye evolutionarily distinct from others of its larger group. An extension of this line of reasoning is that no matter how strange the present animal might appear to us, there must be, buried in its anatomy and biology, features that it inherited from the common ancestor it shared with its closest relative.

When you are beginning an analysis of phylogenetic relationships, you cannot even assume that the group you are working on is in fact a group. Just because someone called a bunch of organisms a group does not mean that they are an evolutionarily real, monophyletic (descending from one source) group. And even in the best of all worlds, you cannot know what will emerge as reflective of the unity of any particular group. Thus, while rodents are set apart by such obviously unique dental features as having teeth that grow throughout the animal's lifetime, another group of mammals may be distinguished by much more subtle dental uniquenesses. Other groups may not be distinctive in their teeth at all, having their uniquenesses perhaps tied to the muscles and bones of their hands and feet. The most one can take as a given about related organisms is that they share some unknown and unpredictable number of biological attributes which they inherited from a common ancestor, and that these features are unique to these organisms because they were unique to, and distinctive of, that common ancestor.

A constant threat to any effort to reconstruct relationships is the possibility that a feature shared by organisms is similar not because of common ancestry, but because it arose through independent and parallel evolutionary events. These similarities are not true, homologous, evolutionary similarities. Features that appear similar, but which proceed through different developmental pathways to get to their final state, are not the same. Apparent similarity in the final, or adult, stage of the developmental process is meaningless unless the developmental foundations are the same.

Theory is great, but in practice it is not always possible to compare organisms other than at their final stages of development. We do not know much at all about the genetic background of most structures. When there are fossils, one is usually fortunate just to have a piece of the extinct organism. One cannot, however, ever lose sight of the possibility that similarity among organisms is due to parallelism, not to common inheritance or homology of structure.

No matter what system, feature, or structure you happen to be studying, the basic criterion for the recognition of homology is similarity. But if similarity can at times be deceiving about its phylogenetic import, it would seem that theories of relatedness must be suspect. There has to be some safeguard, some test, of homology. And there is. A theory of homology can be used to test another theory of homology.

For example, suppose that the relatively thick molar enamel of an orang-utan is homologous with the thick molar enamel of a hominid. I would not want to hang my phylogenetic hat on just this one feature of similarity. However, if other distinctive features limited basically to humans and the orang-utan were to be found, then the suggestion that the thick molar enamel of humans is homologous with the relatively thick molar enamel of the orang-utan would become more viable. As a consequence, the resultant theory of relationships, in this case between humans and the orang-utan, would also become more tenable. In those instances in which more than one possible phylogenetic arrangement can be generated among the same taxa, the theory based on the more robust or more highly corroborated set of homologies would be preferred. But this can be true only if the shared homologies are the right kind of homology.

A demonstration of homology does not specify from which ancestor, of the many ancestors an organism can claim in its evolutionary genealogy, these features were inherited. Some ancestors are more recent and, thus, closer to the origin of descendent taxa. Some ancestors are farther back in time. Homologous features inherited from a more ancient ancestor will be shared by a greater number of related organisms than will homologous features inherited from an ancestor of more recent origin. Thus, while overall similarity between related organisms may indeed be a reflection of the sharing of a number of homologous features, only those homologous features inherited from their *last* or *most recent* common ancestor are reflective of a close relationship between two taxa.

The features that reflect a close relationship are derived features, evolutionary novelties. Features inherited from a less immediate ancestor are primitive retentions. Only evolutionary novelties distinguish one organism from another. Primitive retentions cannot, precisely because they are so broadly distributed among an array of taxa of differing degrees of relatedness. At some level of evolutionary inclusiveness, homologous features are derived and novel, and their possession by a group of organisms reflects the relatedness of these organisms. A derived feature is, however, like the stinging organ of a bee, good only once: it reflects the relatedness of organisms at only one point in their evolutionary past. Once it is "spent" in the ancestry of a group, a derived feature loses its evolutionary significance; it becomes a primitive retention. A used-up derived feature cannot thereafter delineate any other set of relationships, even though it may be retained in hundreds of generations of descendants. In order to delineate smaller and smaller evolutionary or monophyletic groups within the larger assemblage, it is necessary to continue to search for those derived features that would have been distinctive of each successive ancestor.

One might find, for example, that humans and the African apes share fifty features of similarity and that humans and the orang-utan share only forty. But this is only the beginning of the analysis. One is obliged to try to find out how many of these fifty potentially homologous features are actually restricted to only humans and the African apes and, thus, derived features indicative of the phylogenetic closeness of humans and the African apes. Perhaps it is the orang-utan, that will be found to share with humans a greater number of evolutionary novelties. We need to look beyond the uniquenesses that the orang-utan claims only for itself and delineate those uniquenesses it shares with but one or two other taxa in order to understand this primate's place among the hominoids.

5

Primates First, Hominoids Last

ONE OF THE REASONS that generating long lists of features to "prove" the relatedness of taxa has gotten a foothold at all in systematic studies is that there has not been much concern for distinguishing between primitive retentions and those features that really might be reflective of close evolutionary ties. A lot of comparative work has been limited to the organism of special interest and to the handful of organisms that have been agreed upon as being the most closely related to it. But actually, how much of what makes you a complete and functioning organism is unique to you as *Homo sapiens*? How many features are unique to you as a hominid? as a hominoid? as a higher primate, an anthropoid? or even as a primate?

We cannot zero in on the details of our specific relationships if we do not have even an inkling about the relative primitiveness or derivedness of the features we are comparing. We must begin at the beginning, with the very essence of being a primate, and, as if taking apart an artichoke, slough off the ancestral layers of features until we get to the heart of what it means to be a hominoid and what the relationships among the hominoids might be.

The task *sounds* easy — but there is not even general agreement among primate systematists about the taxonomic boundaries of the order Primates. In fact, the question "What is a primate?" continues to be at the center of much controversy and sometimes downright nasty debate.

Prior to turning my research and theoretical interests to human origins, I spent a great deal of time trying to figure out what a primate was and how one could accurately and consistently distinguish a primate from other mammals. I went up my share of evolutionary dead ends, at times discarding theories as quickly as I formulated them, and even came to reject one of the "best" of my earlier "brilliant" ideas.

As William K. Gregory pointed out, Linnaeus "was in the habit of first 'sensing' a natural group and then finding the characters to define it afterward." In the *Systema Naturae,* Linnaeus' major effort to classify the biological world, he came to define the group that would later be known as the order Primates by its members' possession of four incisor teeth in the upper jaw, four incisor teeth in the lower jaw, and one pair of mammary glands confined to the pectoral region of the chest.

In the first edition of the *Systema,* published in 1735, the name Linnaeus gave to this order of mammals was not Primates but Anthropomorpha. Anthropomorpha included the monkeys, whose genus name was *Simia,* as well as *Bradypus,* the sloth. What is most important for the history of systematics and classification is that Linnaeus included in his order Anthropomorpha the genus *Homo.* It was the first time that our species was not set apart taxonomically from all other organisms.

Other taxonomists were outraged at Linnaeus' bold, even heretical, action. How dare he include humans with monkeys, much less sloths! From our vantage point, we would hardly think twice about the reality of our place in nature; what we might find odd is that Linnaeus would group *sloths* with humans and monkeys, since sloths are most like anteaters and armadillos in their broader peculiarities. Sloths don't even have incisors, in either jaw. But sloths do develop only one pair of pectoral mammary glands, and it is the development of only two such glands, rather than two parallel strings of many mammary glands coursing from the chest down along the abdomen, that is distinctive of humans and monkeys.

In 1758, in the tenth edition of the *Systema,* Linnaeus changed the name of the order from Anthropomorpha to Primates, which, in addition to *Homo* and *Simia,* came to claim the genus *Lemur,* a so-called lower primate, as well as the genus *Vespertillio,* a bat. By this time, Linnaeus had changed his thinking about the sloth, which he removed from the order Primates. The reason Linnaeus excluded the sloth from the order and included the lemur and bat can be understood in terms of his expanded definition of Primates. *Simia* now included apes.

For the Linnaeus of 1758, a primate had to have features above and beyond the mere possession of one pair of pectoral mammary glands and four upper and four lower incisor teeth. A primate had to have four lower anterior teeth that were parallel-sided and a pair of somewhat sharp, projecting caninelike teeth in both the upper and lower jaws. Its extremities should end in hands, or in structures that functioned essentially like hands, and its arms were to be separated by clavicles (collarbones). Finally, an animal had to get around most of the time on all fours, climb trees, and eat fruit.

Sloths certainly do not have handlike extremities. Their two- or three-toed extremities are basically hooks. Lemurs most definitely have hands and handlike feet. They are also quadrupedal and climb trees, and their diet contains a goodly amount of fruit. A lemur can be accommodated easily by Linnaeus' new and improved definition of the order.

A bat is another matter altogether. Bats do have sharp, projecting caninelike teeth, two pectoral mammary glands, and clavicles, and some of them do eat fruit. And I suppose one can imagine that a bat's feet qualify as grasping extremities. Nevertheless, Johann Friederick Blumenbach, who is commonly thought of as the father of anthropology, removed *Vespertillio* from the order Primates in his classification of 1779.

Blumenbach's action was probably provoked by equal amounts of science and ego, because he felt as strongly — in a negative way — about Linnaeus' close association of humans with other animals as his British contemporary Thomas Pennant. Pennant, outraged by Linnaeus' classification, had written, "I reject his first division, which he calls Primates or Chiefs of Creation, because my vanity will not suffer me to rank mankind with Apes, Monkeys, lemurs and bats."

After Linnaeus, the delineation of a group of mammals we might call Primates continued in part to be based on aspects of the dentition, but even more crucial was the animal's possession of handlike grasping extremities. Nonhuman primates were frequently classified as Quadrumanes or Quadrumana, which means four-handed. Humans were typically distinguished as Bimanes or Bimana, and the possession of two hands was used time and time again to justify the separation of humans from the rest of the animal kingdom. For example, in his classification of 1800, the French systematist Georges Cuvier identified humans as Les Bimanes because they have thumbs only on the

FIGURE 5A   Various prosimians *(clockwise from lower right)*: the large ruffed lemur, *Varecia variegatus;* the also large but long-legged Verreauxi's sifaka, *Propithecus verreauxi;* the tiniest living primate, the nocturnal mouse lemur, *Microcebus murinus;* a cautious African potto, *Perodicticus potto;* and a bouncy, long-legged bushbaby, *Galago senegalensis.* All of these prosimians have a grooming claw on the second digit of the foot. The potto has a reduced index finger, which is typical of lorises. (The figures are drawn roughly to scale.)

upper extremities. Monkeys and lemurs have "thumbs" on all four "feet," and thus were Les Quadrumanes.

In 1811 the German systematist Carl Illiger used Pollicata instead of Quadrumana to refer to the four-handed primates, but the emphasis and meaning remained the same. Pollicata derives from the Latin word *pollex,* the first digit of the hand. In primates, however, the first digit of the hand is a real thumb, and in especially nonhuman primates, the animal's hands and feet are "all thumbs."

Illiger was one of the few systematists not to invoke two-handedness as the distinguishing characteristic of the genus *Homo.* Instead, he chose to separate humans from the rest of the animal world by their usually erect posture. This physical attribute was reflected in the name Erecta, which Illiger gave to both the family and the order he used to subsume his own species.

A major feature of Illiger's classification has remained with us. It is the term Prosimii, which has come to be equated with the "lower" primates. These days lemurs, lorises, and tarsiers are all thought of as prosimian primates. In 1811, however, Illiger kept *Tarsius* in its own group, Macrotarsi, a name which reflects the animal's development of greatly elongated tarsal bones of the ankle region.

Despite the different emphasis of the occasional Illiger, two-handedness remained for decades a powerful criterion for separating humans from the rest of the animal kingdom. The dogma of two-handedness was so deeply rooted in taxonomists' minds that, even as late as 1863, Thomas Huxley was obliged to devote a large section of his essay "On the Relations of Man to the Lower Animals" to demonstrating that the bones of the human foot were essentially identical to the bones of the human hand. Thus, although there superficially appeared to be a major difference, humans were nonetheless skeletally similar to the more obviously and functionally four-handed primates.

In 1883 the British comparative anatomist William Henry Flower published a classificatory scheme that provided the basic taxonomic breakdown of the order Primates which would persist into the twentieth century. Among other things, Flower undid the taxonomic elevation of our own genus once and for all, firmly reimplanting it within a specific group of mammals, the order Primates. In the almost 150 years of systematics that followed the publication of the first edition of Linnaeus' *Systema,* this was just the second time that humans had

FIGURE 5B Various New and Old World monkeys *(clockwise from lower right):* the African chacma baboon, *Papio ursinus;* the southeast Asian proboscis monkey, *Nasalis larvatus;* the Celebes crested black macaque, *Macaca nigra;* the Abyssinian black and white colobus, *Colobus guereza;* the Amazonian red uakari, *Cacajao rubicundus,* and saddle-backed tamarin, *Saguinus fuscicollis;* and the Latin and Central American prehensile-tailed black spider monkey, *Ateles paniscus.* (Only the New World monkeys — the last three — are drawn roughly to scale.)

been taxonomically united with any other animals. Prior to Flower, only the mid-nineteenth-century French systematist Isidore Geoffroy Saint-Hilaire had been so bold.

Flower divided the order Primates into two major subdivisions, the suborders Lemuroidea and Anthropoidea. Within Lemuroidea, in the family Lemuridae, were the acrobatic lemurs of Madagascar, the tediously slow and cautious lorises of sub-Saharan Africa as well as southeast Asia, and the speedy and bouncy bushbabies, also of sub-Saharan Africa. These days only some of the "true" lemurs of Madagascar are still referred to Lemuridae.

The tarsier as well as the aye-aye were also included in Lemuroidea, but each was relegated to a separate family within this suborder. *Tarsius* was placed in the family Tarsiidae, which eventually came to embrace not only the living tarsier but subsequently discovered tiny fossils thought to be related to the extant form. Since the aye-aye's genus name had not yet been changed to *Daubentonia* from *Chiromys,* Flower coined the family name Chiromyidae for this odd primate.

Flower's grouping of what we might broadly call lemurs, lorises, and tarsiers pulled together those primates that became known as the "lower" primates. Whether one chose to retain Flower's Lemuroidea or to use Illiger's Prosimii as the subordinal name for the entire group, these "lower" primates were thought of as the most primitive of extant primates, and of this group the lemurs were supposed to be the most primitive of all. As a result of this preconception, whenever you wanted to know if another primate was primitive, all you had to do was compare it to a lemur. If the two were alike, the primate you were investigating was, obviously, a primitive primate.

The twentieth-century version of the "great chain of being" was very specific about the relative evolutionary positions of many primates. In fact, a fairly widely accepted notion was that one could justifiably line up the "lower" primates in a perceptible and real sequence of increasing complexity, leading to a group of "higher" primates, whose members, in turn, formed a natural hierarchical series. To many primate systematists, *Tarsius* became the evolutionary link between these seriated groups of lower and higher primates.

The so-called higher primates were the New World monkeys, the Old World monkeys, and the hominoids. The name Anthropoidea, which Flower used to refer to this group of primates, had first been introduced in 1864 by another British anatomist, St. George Mivart. The New World monkeys were thought to consist of two groups: the small, brightly colored, almost birdlike marmosets, and the larger, prehensile-tailed monkeys, exemplified by the typical organ grinder's monkey, which uses its tail like a fifth appendage. The marmosets were placed in the family Hapalidae and the larger forms in the family Cebidae.

Just as the lemurs have been regarded as the most primitive of prosimian primates, the New World monkeys have traditionally been perceived as being the most primitive of the anthropoids. The evolutionary chain was thought to project from *Tarsius* into Anthropoidea via the marmosets and then the cebids. Today, there is still much to learn about New World monkey systematics. But a former student of mine, Susan Ford, has been among those who have argued quite persuasively that the marmosets, referred these days to the family Callithricidae, are actually specialized rather than primitive types of New World monkeys. For instance, marmosets are unique in that, as a rule, they give birth to twins; having single births is the more common anthropoid condition.

The Old World monkeys, placed then as well as now in the family Cercopithecidae, have historically been the least problematic group. Although they are quite geographically widespread, ranging from Africa to northern Japan, and live in a wide variety of habitats, from the arid plains of Ethiopia to the dense tropical rain forests of southeast Asia, the Old World monkeys are perhaps most easily compartmentalized because of their dental specializations.

Regardless of the animal, whether it is a savanna-dwelling baboon or an arboreal *Colobus,* an Old World monkey's cheek teeth are most distinctive in that the cusps are lined up in a series of crosswise ridges. A premolar has only one ridge, or loph, coursing across it, linking the tooth's two cusps. The four cusps of each molar are lined up in pairs so that there are two lophs on each tooth, each loph oriented from the tongue to the cheek side of the tooth. With two lophs per molar, and with molars typically being the paleontologist's favorite taxonomic yardstick, the configuration is called bilophodonty. Since bilophodonty is so rarely seen — not just among primates but among mammals in general — it is reasonable to conclude that it is a derived dental state. Among primates, then, the presence of bilophodonty would seem to attest to the unity of those taxa which possess it. And the most blatantly bilophodont primates are the Old World monkeys, the cercopithecids.

FIGURE 5C   A left upper second molar of a typical Old World monkey *(left)* and an orang-utan *(right)* viewed from the tongue side of the tooth, looking across the chewing surface of the tooth. Old World monkeys are distinguished as a group by the ridges *(arrows)* that connect opposing pairs of tongue and cheek cusps on their molars. This configuration is called "bilophodonty." Hominoids retain separate, unaligned molar cusps. Since the upper and lower molars of an Old World monkey are bilophodont, these teeth look like mirror images of each other.

Flower did not refer to the apes and humans as hominoids. Rather, the apes were simply allocated to the family Simiidae. Humans, of course, were Hominidae.

But what was the taxonomic common denominator for the entire assemblage of primates? What held the order Primates together?

The features crucial to being a primate had been outlined in 1873 by St. George Mivart. In his landmark publication "On *Lepilemur* and *Cheirogaleus* and on the Zoological Rank of the Lemuroidea," Mivart presented a definition of the order that came to dominate as well as to confound and confuse primate systematics for decades. Mivart's definition of the order Primates was:

> Unguiculate claviculate mammals, with orbits encircled by bone, three kinds of teeth, at least at one time in life; brain always with a posterior lobe and calcarine fissure; the innermost digits of at least one pair of extremities opposable; hallux with a flat nail or none; a well-developed caecum; penis pendulous; testes scrotal; always two pectoral mammae.

Some of these characteristics had been used by earlier systematists. In the first edition of the *Systema Naturae,* Linnaeus had defined a primate as a mammal that possessed only one pair of pectoral mammary glands. In the tenth edition, Linnaeus added the criterion of having a clavicle, which is what Mivart's "claviculate" refers to. Although Linnaeus had also included in his definition of the primates the possession of caninelike teeth as well as the possession of incisor teeth anteriorly in the jaw, it was Cuvier who had specified that a primate must have three types of teeth.

In addition to their teeth, Cuvier had proposed that groups of animals could be distinguished by differences in the anatomy of the tips of their digits. There were hoofed animals and those whose fingers and toes bore nails. Mivart's term "unguiculate" refers to the latter condition, bearing nails. However, Cuvier did not make a distinction between those nonhoofed animals with compressed nails or claws, such as a cat or a squirrel, and those nonhoofed animals with flattened nails, such as a baboon or a lemur. In this regard, Mivart's specific reference to the hallux, or big toe, with a flat nail is significant. In fact, the issue of whether the ancestral primate was truly clawed or whether its fingertips bore nails has been and still is a hotly debated topic.

The reference in Mivart's definition of the order Primates to "the innermost digits of at least one pair of extremities opposable" reflects an understanding of the structural similarities between the human hand and foot that only Huxley and possibly Illiger before him had appreciated. Thus, while one could easily continue to group most primates together on the basis of their possession of four "hands," Mivart had to limit the possession of a thumblike digit to one pair of limbs so that humans could be accommodated as well.

Mivart's interest in the brain as a potential source of taxonomic and systematic information continued a tradition that his contemporaries Huxley and Haeckel had also inherited. The features of the brain and nervous system were first employed systematically by the French comparative anatomist Henri Marie Ducrotay de Blainville in his influential 1816 classification of mammals. The classifying of mammals on the basis of their cerebral development was carried further by Prince Charles Lucien Bonaparte, who in 1837 divided the mammals into two groups. One group, with brain surfaces that must be reflective of at least some intelligence, Bonaparte called Educabilia. The second and obviously inferior group he referred to as Ineducabilia. Bonaparte's Educabilia consisted of the carnivores, ungulates, manatees and other sirenians, whales, and of course primates. This group was characterized by having the large portion of the brain, the cerebrum, subdivided by a crease or fissure into two or three lobe-like "segments." The Ineducabilia, which included the sloths, bats, insectivores, and rodents, had only a single-lobed, undifferentiated cerebrum.

We might perhaps trace Sir Arthur Keith's notion of a "Cerebral Rubicon" — that brainy barrier between humans and the rest of the organic world — to these earlier brain-based classifications. But these classifications were first distilled and filtered though the brain of Sir Richard Owen, one of the most powerful and influential deans of British zoology. Owen, an ardent anti-Darwinian and anti-evolutionist, became one of Huxley's most vocal adversaries. In 1868, Owen transformed Prince Bonaparte's division of Educabilia and Ineducabilia into his mammalian "subclasses" Gyrencephala (a gyrus is a convolution, a fold, in the surface of the brain) and Lissencephala (referring to an unconvoluted brain). Humans were promoted to their own subclass, Archencephala, the "chiefs of the brain," because of their large brains with their greatly convoluted surfaces.

Mivart disagreed with Owen's reasons for the separation of humans from other animals, and especially from other primates. Indeed, Mivart felt that some groups of animals have consistent patterns to the grooves and folds of their brain's surface and to its external architecture. In particular, Mivart concluded that the brain of a primate always has a posterior lobe with a particular crease, called the calcarine fissure.

How workable was Mivart's definition of the order Primates? Since so many of the features listed are aspects of soft-tissue anatomy, it is doubtful that any fossil could be identified as a primate. But even if we focused solely on extant mammals, the total definition is not workable. Most of the characters in Mivart's list are not specific to just one group of mammal. For example, the possession of three kinds of teeth — incisors, canines, and the premolars and molars that together are the cheek teeth — is not limited to one subset of the class Mammalia. Some marsupials, like the opossum, and some placental mammals, like the pig or bear, develop three different classes of teeth. Most mammals have a well-developed caecum, or blind-pouched portion of their intestinal tract. In carnivores, such as cats, the testes descend into a scrotum. The development of only two pectoral mammary glands is found not only, as we have already discussed, in the sloth and the bat, but also in the so-called flying lemur (a "winged," gliding mammal whose face is reminiscent of a lemur's). Tree shrews have a posterior cerebral lobe with a calcarine fissure toward its midline. And if there truly is a group we might call Primates, it is not the only collection of mammals to possess a clavicle. The clavicle may be longer in primates than in other claviculate animals, but possessing one is not restricted to primates. In fact, even reptiles develop a clavicle.

Those features from Mivart's list of primate attributes that remain as being truly primate are few indeed. One of them is the pendulous penis. Other mammals have the penis either pulled in abdominally or tethered longitudinally to the external abdominal wall. The development of flattened nails, or at least a flattened nail on the big toe, is also distinctive among mammals. Even the aye-aye, with its compressed, clawlike nails (or maybe they are just claws), has a flattened nail on its big toe.

An opposable thumb or big toe is something special. Such a digit on each extremity gives the fortunate animal not just four "hands," but four *grasping* "hands." The possession of an opposable thumb or big toe may not be restricted to primates, but it is not all that common among mammals either. While many mammals have five-digited extremities, few animals have innermost digits capable of extending laterally away from the other digits. Aside from prosimians and anthropoids, a grasping foot — that is, an opposable first digit — is found only among some marsupials. An example is the American opossum, which has an opposable big toe. Thus, even though these attributes of the hand and foot are not entirely restricted to one collection of animals, they are not so common that they cannot be brought to bear on the question of an animal's primateness.

Mivart's reference to the development of a bony strut along the side of the eye, which thus encircles the orbit (the eye socket) with bone, also deserves specific comment. Most mammals do not have any bony protection on the side of the eyeball. Instead of bone, the side of the eye of these animals is buttressed by some of the heavy chewing muscles that move the mandible. These muscles, the temporal muscles, emanate from the uppermost tip of the ascending portion of the mandible. They course behind the cheek bone and then fan out along the side of the skull, covering in part the temporal bone, which contributes to the cranial wall. While chewing, or just moving your jaw, you can trace the extent of our own temporal muscles with your fingertips. You will feel the edge of the muscle just behind the bony strut on the side of our eye, arcing backward over the side of your skull and then descending toward your ear.

In most mammals, the temporal muscle alone buttresses and protects the eye. In other mammals, most notably cats, there is variable growth of a bony buttress. In some cats, bone descends from the region above the eye and bone ascends from below the eye, but these two "branches" do not meet to form a complete bar. In other cats, a complete postorbital bar is formed.

Complete postorbital bars are typical of horses as well as of the enigmatic tree shrew. Indeed, the development in the tree shrew of a complete postorbital bar was for years taken as almost positive proof that this nervous and frenetic mammal is a primate, if not even the living representation of the ancestral primate. Elsewhere among mammals, postorbital bars are typical of the prosimians (lemurs, lorises, and *Tarsius*) and the anthropoids (New World monkeys, Old World monkeys, and hominoids).

In lemurs and lorises, but to a much lesser degree in the tarsier, the postorbital bar is just that: a bar or strut. If you had a lemur skull in front of you, you could hook your finger around this bony strut. The orbital region of an anthropoid, like that of a lemur or loris, is protected laterally by a strut of bone. But anthropoids are distinctly different from lemurs and lorises in that the growth of neighboring cranial bones almost completely incorporates the postorbital bar, thus blocking off the whole side, as well as almost all of the back, of the eye. Aside from the occasional hole through which an artery or nerve might course, an anthropoid's orbital region is essentially a bony socket. Thus, whatever we might conclude about the evolutionary significance of developing a complete, but strutlike, postorbital bar, it would certainly appear that the development of postorbital closure is significant and unites those primates known as Anthropoidea.

Aside from a pendulous penis and a flattened nail on the big toe, there aren't any characteristics in Mivart's list that are restricted to one, and only one, group of mammals. Indeed, if one were to try to determine whether a given fossil was a primate on the basis of these two remaining elements, one would be hard put to substantiate anything but the presence of a flattened nail — assuming that you could find a fossilized big toe, which, because of the general lightness and small size of the bone, is one of the least frequently found parts of the body.

In general, systematists have had difficulty defining the order Primates on the basis of features restricted to this group. There has been a tendency to swamp the definition of the order with long lists of traits in the hope of eliminating the confusing and taxonomically troublesome mammals by their eventual failure to satisfy all the criteria necessary to qualify as primates. Given Mivart's attempted definition of the order and the attempts that followed, it is perhaps no wonder that an apparently frustrated F. Wood Jones would comment in 1929 that "there is no single character . . . which constitutes a peculiarity of the Primates; for a primate animal may only be diagnosed by possessing an aggregate of them all."

This does not make evolutionary sense, however. Surely the ancestor of a subsequently successful group — as primates appear to be — would have been distinct from other organisms in at least one derived, evolutionarily novel feature that it claimed for itself. Nevertheless, and essentially for not very good systematic reasons, it became almost fashionable in the twentieth century to declare that the order Primates, unlike other mammals, could not be defined by concrete and discrete characteristics. Thus, contrary to the very essence of evolutionary prediction — that organisms become distinct through the acquisition of evolutionary novelties — primates were supposed to be distinguished by their *lack* of uniquenesses.

A leading advocate of this systematically negative view of the distinctiveness of the order Primates was the British anatomist and primatologist Sir Wilfred Le Gros Clark, who reigned supreme over the field of primatology until his death in the early 1970s. In one of his most frequently cited works, Sir Wilfred wrote:

The order Primates consists to-day of what seems to be a rather heterogeneous collection of types. In fact, it is not easy to give a very clear-cut definition of the order as a whole, for its various members represent so many different levels of evolutionary development and there is no single distinguishing feature which characterizes them all. Further, while many other mammalian orders can be defined by conspicuous specializations of a positive kind which readily mark them off from one another, the Primates as a whole have preserved rather a generalized anatomy and, if anything, are to be mainly distinguished from other orders by a negative feature — their lack of specialization. This lack of somatic specialization has been associated with an increased efficiency of the controlling mechanisms of the brain, and for this reason it has had the advantage of permitting a high degree of functional plasticity. Thus, it has been said that one of the outstanding features of Primate evolution has been not so much progressive *adaption* (as in other mammalian orders) but progressive *adaptability*.

Le Gros Clark's sentiment is not derived solely from his nineteenth-century predecessors. George Gaylord Simpson, who for decades wielded a heavy hand over American vertebrate paleontology, had failed in his 1945 magnum opus on the classification of the mammals to provide a workable definition of primates. Seemingly never at a loss for definitive systematic statements, Simpson could have been expected to have made major inroads into the problem of defining the order Primates. But this, or any real clarification, was not forthcoming:

> The primates are inevitably the most interesting of mammals to an ego-centric species that belongs to this order. No other mammals have been studied in such detail, yet from a taxonomic view this cannot be considered the best-known order, and there is perhaps less agreement as to its classification than for most other orders. A major reason for this confusion is that much of the work on primates has been done by students who had no experience in taxonomy and who were completely incompetent to enter this field, however competent they may have been in other respects, and yet once their work is in print it becomes necessary to take cognizance of it . . . The peculiar fascination of the primates and their publicity value have almost taken the order out of the hands of sober and conservative mammalogists and have kept, and do keep, its taxonomy in turmoil. Moreover, even mammalogists who might be entirely conservative in dealing, say, with rats are likely to lose a sense of perspective when they come to the primates, and many studies of this order are covertly or overtly emotional.

The feeling conveyed here — not just anger toward those who study primates but annoyance at having to deal with these primatologists and their studies — is typical of Simpson and of many other paleontologists and systematists as well. Nevertheless, it is the case that the potential glamour of solving some riddle about human origins has attracted many scientists who, while excellent technicians in their own primary fields of research, are totally unappreciative of the basic tenets of method and theory in evolutionary biology. (To complicate the problem, there are also disagreements among the dyed-in-the-wool systematists about which method is appropriate and which model best explains the evolutionary process.)

Basically, people who study rats do not get the same attention as a Leakey or a Johanson. But because of the chanciness of making spectacular fossil discoveries, there will be a limited number of Leakeys and Johansons, and, thus, as one might expect, jealousies exist among primatologists, perhaps especially among those striving to succeed in paleoanthropology. As Matt Cartmill, an ardent student of

FIGURE 5D Various mammals considered, from time to time, to be members of the order Insectivora *(from top, clockwise):* the southeast Asian tree shrew *(Tupaia glis);* the East African long-eared elephant shrew *(Elephantulus rufescens);* a "proper" common shrew *(Sorex sorex);* the European hedgehog *(Erinaceus europeus);* and a so-called flying lemur *(Cynocephalus volans,* shown less than scale). Only the common shrew and hedgehog are related and recognized as true insectivores.

"lower" primates with a flair for incisive reductionism, has recently put it, "From the time of Darwin and Huxley to the present, the wish to understand man's place in nature has drawn scientists in disproportionate numbers to the study of the relatively small and unsuccessful mammalian order Primates — an order that is, after all, about equal in size and adaptive diversity to the single family of squirrels."

After venting his spleen about the "state of affairs" in primate systematics and having made his opinion of primate systematists perfectly clear, Simpson does provide a brief outline of his thoughts on primate origins and the general course of subsequent evolution of the group. No doubt influenced in part by Thomas Huxley's 1880 treatise on the taxonomic arrangement of the mammals and by Le Gros Clark's early twentieth-century comparative studies of the brain, Simpson begins by deriving the ancestral primate from another group of mammals, the insectivores.

Who are the insectivores? Good question. The "group" Insectivora has taken such systematic abuse that, at one time or another, an untold number of mammals have been relegated to this order.

"Insectivores" are supposed to be primitive, generalized mammals from which more advanced or specialized mammals could have arisen. But mammals whose affinities could not be adequately assessed were also commonly dumped into the order Insectivora. The upshot was that, with enough animals thrown into a taxonomic wastebasket that by definition was an orphanage for primitive, unclaimed mammals, the origin of virtually any other mammalian group could be sought from within Insectivora.

Ernst Haeckel was the first to introduce the concept of an insectivore into systematics. His Insectivora of a century ago was restricted primarily to hedgehogs and shrews. Subsequently, however, the tree shrews (which have nothing to do with proper shrews), the so-called flying lemurs, and the puzzling "elephant" shrews (again, having nothing to do with proper shrews, or tree shrews, or even elephants, except that they have long snouts) were assigned to Insectivora — basically because nobody knew what to do with them. Attempts at resolving the problematic affinities of these mammals are only now being made. Tree shrews, flying lemurs, and elephant shrews may in fact represent separate and distinct types of mammals, and the "real" insectivores may be Haeckel's original members, the hedgehogs and the proper shrews.

But the endeavor to resolve the problems does not necessarily mean that the use of Insectivora as a taxonomic wastebasket will cease. Taxonomic wastebaskets are useful conveniences. They can receive the messy organisms, those whose removal from your group of interest tidies up that group — even if such an action adds to the taxonomic clutter of another group.

Plesiadapiforms are one example of a group threatened with exile to Insectivora. Plesiadapiforms are fossil mammals from quite ancient Paleocene deposits, some older than sixty million years, of Europe and North America. For decades now, especially because of George Gaylord Simpson's papers of the 1930s, plesiadapiforms have been characterized as dental primates. That is, in the shape and configurations of their cheek teeth, plesiadapiforms can be distinguished from many other mammals in the same ways in which animals everyone accepts as being primates can also be distinguished among mammals.

There are some very basic dental features — having cusps that are not too tall and pointed, and having molars that are noticeably "squared up" on their sides — that characterize plesiadapiforms as well as those animals we would agree represent living primates. However, because we know from a few skulls and preserved hand and foot bones that some plesiadapiforms did not have a suite of *additional* features found in living primates, the extinct forms have been degraded by some primate systematists as being not quite primate enough. Thus, in order to construct a group we might call Primates that can be defined on the basis of more than just dental features, these primatologists have campaigned quite forcefully for discarding the plesiadapiforms into the insectivore wastebasket.

The removal of the plesiadapiforms from the order Primates may bring some apparent order to Primates, but it violates the evolutionary integrity of another group, Insectivora. I have never understood why being dentally primate was not enough to be reflective of the unity of a distinct and evolutionarily real group within Mammalia. A large part of the problem may be that there is a tendency toward preconceived notions of just how much it takes to cross the "Primate Rubicon," and how much more it takes to remain there.

The idea of crossing some boundary between being a presumed insectivore ancestor and being an early, primitive primate descendant derives in part from adhering to the notion that the order Primates originated via a major adaptive shift. Le Gros Clark formalized this transformationist viewpoint as a series of "prevailing evolutionary trends" in primate evolution.

A primate's link to the most primitive of insectivores, which Simpson took to be the hedgehog, was the tree shrew. Especially because of Le Gros Clark's perception of similarities between tree shrews (the tupaioids) and primates in brain configuration, Simpson and others came to believe that "the tupaioids are either the most primate-like insectivores or the most insectivore-like primates." Just as Raymond Dart's *Australopithecus* became the apes' "manlike" link with Eugène Dubois's "apelike man" link to modern humans, the tree shrew was believed to represent the missing link in the evolutionary chain of being between primates and the rest of the mammalian world.

With an emphasis on the extreme primitiveness of primates, Le Gros Clark enumerated nine important aspects of being a primate. These features of "primateness" built upon a progressive tendency within the order to depart from primitiveness through the development of increasing anatomical and behavioral complexity. "Broadly speaking," Le Gros Clark wrote,

the order Primates can be defined as a natural group of mammals distinguished by the following prevailing evolutionary trends:

1. The preservation of a generalized structure of the limbs with a primitive pentadactyly [the possession of five digits on hand or foot], and the retention of certain elements of the limb skeleton (such as the clavicle) which tend to be reduced or to disappear in some groups of mammals.
2. An enhancement of the free mobility of the digits, especially the thumb and the big toe (which are used for grasping purposes).
3. The replacement of sharp compressed claws by flattened nails, associated with the development of highly sensitive tactile pads on the digits.
4. The progressive abbreviation of the snout or muzzle.
5. The elaboration and perfection of the visual apparatus with the development of varying degrees of binocular vision.
6. Reduction of the apparatus of smell.
7. The loss of certain elements of the primitive mammalian dentition, and the preservation of a simple cusp pattern of the molar teeth.
8. Progressive expansion and elaboration of the brain, affecting predominantly the cerebral cortex and its dependencies.
9. Progressive and increasingly efficient development of those gestational processes concerned with the nourishment of the foetus before birth.

By anyone's standards, Le Gros Clark's "trends of primateness" were a smashing systematic success. They were repeated and summarized in virtually every text and major reference work on the order. But aside from painting what appears to be a nice picture of evolutionary transformation and progress at work, Le Gros Clark's "evolutionary trends" don't seem to me to be at all useful.

How, for instance, could you decide whether any organism, fossil or extant, was a primate in the first place? You couldn't. Le Gros Clark's "trends" exist only as a result of having first lined up those mammals that one thinks are primates and then "discovering" what is supposedly new and different and primate about them. One has a starting point in the tree shrew, an end point in *Homo sapiens,* and a bunch of lemurs and lorises and monkeys to play around with in between.

Trends do not, and cannot, define an evolutionary group, because real evolutionary groups have boundaries, almost tangible edges. Their discreteness in nature derives from the fact that their founding ancestors must themselves have been distinct entities within the natural world. The search for trends in nature is nothing more than nineteenth-century transformationalism — again, an evolutionary version of the great chain of being — pursued either throughout a hierarchical arrangement of living forms, as in Le Gros Clark's trends, or in ascension through the fossil record to the present, as in so many paleonanthropological analyses.

If we eliminate from Le Gros Clark's list the elements of progressive trends in primate evolution, we are left with some "hard" morphologies, but they are of the wrong sort because they emphasize the primitive and not the evolutionarily novel. For instance, the possession of a clavicle, as we saw in discussing Mivart's definition of the order Primates, is not unique to primates. And if a particular skeletal configuration is widespread enough among animals to be referred to as a "generalized" skeleton, its possession by any particular mammal can hardly be construed as being of any phylogenetic significance.

It is perhaps ironic that the heir to Le Gros Clark's British primatological throne, Robert D. Martin, until recently of University College, London, had no time for these trends of primate evolution. In one of his earliest papers on the problem of defining a primate, Martin went right to the heart of the matter, pointing out that these trends are "largely an artificial feature developed in the minds of the 'highest' primates, which picture a line of progressive evolution ultimately leading to themselves."

It must seem almost unbelievable that systematists could delineate with relative ease the morphological essences of being an ungulate or a carnivore, for example, but could not arrive at any reasonable working definition of the order Primates. Perhaps this is because the order, Linnaeus' "Chiefs of Creation," contains the chief of the chiefs, and thus an unduly demanding, unrealistic set of taxonomic criteria might have been established. But even if we continue to conceive of ourselves as being somehow different from other organisms and thus exempt from the whims of evolution, all other primates are still just plain old animals.

Among the features that have been proposed over the years as being definitely "primate," those which keep cropping up are grasping hands and feet. This does seem reasonable since, aside from some marsupials, no other mammals have these physical attributes. And since only a few marsupials have grasping hands and feet — and marsupials are otherwise unique, for instance, in their development of four, rather than three, molars, as well as in unusual features of their reproductive physiology (the fetus leaves the womb to continue its development in its mother's pouch) — it would seem that the marsupials with grasping hands and feet have distinguished themselves further among their group.

Prosimians and nonhuman anthropoids have divergent thumbs and big toes and a big toe that bears a flattened nail on its tip. As was recognized well over a century ago by St. George Mivart, the possession of a flattened nail on the first digit of the foot is unquestionably unique among mammals.

In addition to these aspects of their hands and feet, prosimians and anthropoids develop a postorbital bar. And the development of a complete bony strut alongside the eye is truly an uncommon feature among mammals. Aside from horses and the occasional cat, the development of postorbital bars is most characteristic of those mammals with grasping hands and feet with a flattened nail on the big toe.

While some marsupials might share with prosimians and anthropoids the possession of an opposable big toe, and some other mammals might share with prosimians and anthropoids the development of a postorbital bar, only prosimians and anthropoids share both of these features as well as the flattened nail on the big toe. Thus, it would seem that there are features that do distinguish a subset within Mammalia, a group that includes our own species. If we had no knowledge of a primate fossil record, we would certainly feel justified in delineating a group of animals within Mammalia on the basis of these three characteristics. These animals would be distinguished further among mammals by their having cheek teeth whose cusps are lower and bulbous rather than tall and pointed and by their having somewhat squared-up molars. Thus, on the basis of uniquenesses of their teeth, their extremities, and their skull, a group of mammals can be defined.

But there are fossils relevant to the history of primates, and they, too, must be taken into account. We do know of some fossil material as old as perhaps fifty-five million years — earliest Eocene deposits — from North America as well as Europe which can be brought to bear on the issue of being primate. There are specimens that, in preserved dental morphology, facial configuration, and the disposition of the digits of the hand and foot, conform to the broad picture of primateness. These Eocene fossils include early possible representatives of lemurs, lorises, and tarsiers. And, if Ian Tattersall, curator of physical anthroplogy at the American Museum of Natural History and lemurologist extraordinaire, and I are correct in our analyses, some of these fossils actually represent ancient species whose closest relatives are to be found among the living prosimians.

The close association of Eocene fossils with specific modern prosimians has profound implications for trying to understand the antiquity not only of lemurs and lorises but of the entire order. Taking the relative positions of the continents into consideration with the presently known geographic distribution of fossil and living prosimians, Tattersall and I have argued that the ancestral primate must have become differentiated among mammals well before the beginning of the Paleocene epoch — prior to sixty-five or seventy million years ago, before the demise of the dinosaurs. If deposits as early as the Eocene contain identifiable relatives of specific living prosimians, then it should not come as a surprise to learn that the geological epochs that followed the Eocene have to a large extent yielded fossils with affinities to other groups of extant primates.

Elwyn Simons has for years believed that the possible ancestor of the Old World monkeys and the possible ancestor of the hominoids have both been discovered in the post-Eocene (that is, Oligocene) deposits of the Fayum region of Egypt, some forty kilometers outside Cairo. Since this is the only Old World Oligocene deposit to yield fossil primates, I am highly skeptical of the idea that the Fayum, a mere speck on the map, was home to the ancestral populations of both of these major primate groups. On morphological grounds, I think Simons's ancestral Old World monkey is so enigmatic that the best one can say is that it is an anthropoid. A preserved skull of one such genus, *Apidium,* does reveal that the animal had some degree of postorbital closure — but that is a feature of any and all anthropoid primates, not just Old World monkeys.

The Fayum's supposed early hominoid, called *Oligopithecus,* the ape of the Oligocene, may not even be an anthropoid primate, much less an ancestor of a small group of anthropoids. Philip Gingerich of the University of Michigan, one of Simons's earlier students, has suggested that *Oligopithecus* is an evolutionary holdover of one of those Eocene fossil prosimians. My recent review of primate evolution and systematics resulted in an even more radical conclusion: that the affinities of *Oligopithecus* lie more precisely with the modern lemur group of Madagascar.

Subsequent to the Oligocene, in Miocene deposits of South America as well as Europe, Africa, and farther east, fossils reminiscent of modern groups are the rule. In South America, some of the fossils are actually relatable to certain living genera of New World monkeys. Especially from Europe but also from the Mediterranean countries of Africa have come fossils that are unquestionably Old World monkeys, anthropoids with bilophodont molars. And of course there are the possible ape-related forms, such as *Dryopithecus, Proconsul,* and *Limnopithecus,* and the more orang-utan-like *Sivapithecus* and *Gigantopithecus.*

Granted, this is a rough and admittedly speedy review of most of primate paleontology. But, to be blunt, if a fossil can be associated with a specific prosimian or anthropoid or is at least obviously related to a modern group of primates, the animal is unquestionably a primate. If a fossil form looks pretty much like a lemur or a baboon, then that animal must, in its makeup, have not only the peculiarities of a lemur or a baboon, but those novelties that ultimately unite lemurs or baboons as larger evolutionary groups. Thus, at this point, and especially with regard to the ultimate question of what is a primate, those fossils whose specific phylogenetic relationships can be resolved with some certainty are not only not interesting, they are not enlightening.

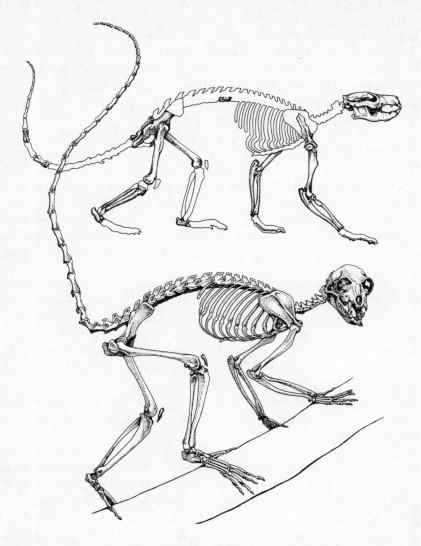

FIGURE 5E The reconstructed skeleton of the Paleocene fossil dental primate
Plesiadapis (*above*, with the best-known skeletal elements stippled) compared
with that of *Lemur*, a so-called primate of modern aspect. *Plesiadapis* is primate
in its molars, but the lemur is "more" primate in having a postorbital bar protecting the side of its eye and in having grasping hands and feet. A specialization
of *Plesiadapis* is its large anterior teeth. A specialization of a lemur is its comblike
lower anterior teeth.

Are there any fossils available that will provide a test of our notions of primateness? that will challenge our preconceived notions?

Of course there are.

They are the Plesiadapiformes, those ancient fossils that Simpson, for example, thought were primates because of their somewhat low-cusped cheek teeth and their rather squared-up molars. In their teeth, plesiadapiforms are definitely dentally primate.

There are a plethora of plesiadapiform genera and species so far known — primarily from Paleocene deposits of North America and Europe, but also from Eocene as well as even Oligocene sites. However, most of our knowledge of the group comes from fragmentary jaws with teeth, or just isolated teeth. If we were content to accept features of the dentition as the common denominator of being primate, the plesiadapiforms, the prosimians, and the anthropoids, as well as any fossil forms whose affinities clearly lie with any of these taxa in particular, would belong to the group. However, being dentally primate has not been quite good enough for plesiadapiforms to qualify as primates, in the judgment of some systematists.

The skulls of some plesiadapiforms are known. The genus for which the most cranial material is known is *Plesiadapis,* whose name is reflected in the name of the larger group, Plesiadapiformes. We have discovered from studying the cranial remains of *Plesiadapis* and a few other genera that these plesiadapiforms apparently lacked a postorbital bar. It has become popular to conclude that if the genera and species referred to Plesiadapiformes all belong there and all of these taxa did share a common ancestor, no plesiadapiform possessed a postorbital bar. I will not take issue here with these assumptions.

In terms of our knowledge of the postcranial skeleton of the plesiadapiforms, the genus *Plesiadapis* is once again the best known of the group. The good fortunes of preservation have provided virtually unquestionable associations of limb bones with jaws and teeth of this plesiadapiform.

It appears that *Plesiadapis,* at least, had neither grasping hands and feet nor flattened nails on any of its digits, as it would if it were a "true" primate. Instead, the squirrellike *Plesiadapis* was similar in its hands and feet to a primitive mammal in the disposition of its toes and fingers and in the possession of compressed, "fissured" claws, which it no doubt used for scurrying up and down the trunks of trees.

A common element of the typical stories of primate evolution is that there was a shift from a reliance on the senses of the snout to that of vision. The emphasis on sight is reflected in the relatively larger orbits of an extant primate and in the development of a postorbital bar. Since the skulls of *Plesiadapis* and other plesiadapiforms have small orbits that were not encircled by bone, these animals, and thus plesiadapiforms in general, had not "shifted" to a primatelike, visually oriented lifestyle. Not only were the orbits of a plesiadapiform small and lacking a postorbital bar, there is also evidence that the animal relied heavily on highly sensitive whiskers, or vibrissae, to get around.

All mammals have a particular facial canal, called the infraorbital canal because it passes below the orbit and penetrates the cheek bone just below the rim of the orbit. The sensory infraorbital nerve courses along the canal and emerges through the cheek bone to send branches to areas of the lower face. A survey among mammals of the size and associated structures of this nerve indicates that a large nerve trunk is correlated with the development of whiskers and their sensitivity to touch. Matt Cartmill and Rich Kay of Duke University discovered that the skull of the plesiadapiform *Palaechthon* preserved a very large infraorbital canal and facial opening, which obviously meant that the animal was definitely more snout- than eye-oriented.

From what we can piece together, *Plesiadapis, Palaechthon,* and their closest evolutionary allies were not highly arboreal nor were they visually and manually dominant. Rather, these little animals were probably like shrews — small-eyed, heavily whiskered animals, darting through the underbrush guided by a dominant sense of smell.

But the plesiadapiforms were different in one major way from shrews, hedgehogs, or any other animal that might have been around in the Paleocene: they had cheek teeth resembling those of modern primates. Thus, even if one does not care to call them primates, plesiadapiforms were the first of our broader evolutionary group. Although our teeth are certainly different from those of *Plesiadapis,* we share with it a basic pattern that in all likelihood would have characterized the ancestral primate and distinguished it from the rest of the mammalian world. This is our primate heritage. Our dental pattern saw its origins at least seventy million years ago.

Most of our skeletal and cranial elements were also inherited from this ancestor, but other than in the details of its teeth, there was nothing particularly primatelike about this ancestor. As emphasized

A young Bornean orang-utan (P. Rodman)

An adult male Sumatran orang-utan. Compared to Bornean orang-utans, Sumatran orang-utans have more "feathery" hair, and the males have "mustaches," "beards," and less pronounced cheek flanges. (H. Rijksen)

An adult female Sumatran orang-utan with child. As in humans, the breasts of orang-utans are broadly separated. (H. Rijksen)

An adult female Bornean orang-utan with child (J. Mitani)

An adult male Bornean orang-utan with the characteristically large, squared, protruding cheek flanges (P. Rodman)

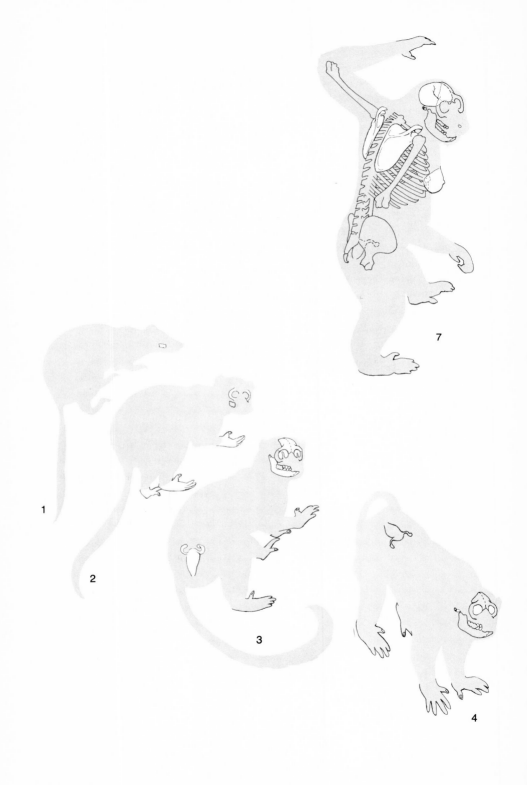

1

2

3

4

7

Hypothesized ancestors of various primate groups. Each successive ancestor comprises features of differing evolutionary significance: derived features (highlighted), which make it distinct, and primitive (outlined in black), which are inherited from ancestors that came before it. The ancestral primate (1) would have been distinguished especially by the shapes of its molar teeth; the ancestor of prosimians (lemurs and lorises) and anthropoids (New and Old World monkeys, apes, and humans) (2) by the development of grasping hands and feet, a flattened nail on the big toe, and a bony bar protecting the sides of the eye; the ancestor of anthropoids (3) by reduction in the number of premolars, early fusion of the frontal bone of the skull and of the two halves of the lower jaw, and the development of a single-chambered uterus; the ancestor of Old World monkeys and hominoids (gibbons, the large apes, and humans) (4) by further reduction in premolar number and changes in the bony ear region, such as the development of a bony tube; the ancestor of hominoids (5) by changes in the shoulder region, upper arm, rib cage, and pelvic girdle; the ancestor of the large hominoids (chimpanzees, gorillas, orang-utans, and humans) (6) by further alterations of the shoulder and pelvic regions; and the ancestor of humans and orang-utans (7) by the development of marked asymmetries of the brain, separation of the mammary glands, and further alterations of the shoulder region. (J. C. Anderton)

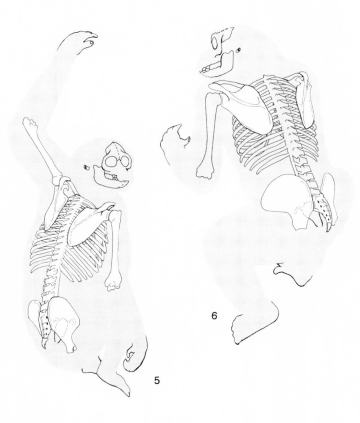

5

6

VARIOUS PROSIMIANS: (1) A sifaka (Madagascar), noted for its long legs and tail (R. Sussman). (2) A tarsier (Southeast Asia) (D. Haring). (3) A ring-tailed lemur (Madagascar) (D. Haring). (4) A gentle lemur (Madagascar) chewing on its primary food source, a bamboo shoot. Note the "handlike" grasping feet and the flattened nail on the big toe. (D. Haring)

3   4

VARIOUS NEW AND OLD WORLD MONKEYS: (1) A widow monkey (South America) feeding on berries (W. Kinzey). (2) An emperor tamarin (South America) showing off its long and wispy "mustache" and "beard" (W. Kinzey). (3) A savanna-dwelling baboon (sub-Saharan Africa) (S. Gaulin). (4) A spider monkey (South America) hanging by only one arm and its prehensile tail (P. Rodman)

Above: An adult male Bornean orang-utan (throat sac and cheek flanges fully developed) shuffling along the ground, its long fingers and toes curled into "fists" so as not to trip over them (G. Raab).
Below: A tropical rain forest (Borneo) — home of the orang-utan (P. Rodman)

by Le Gros Clark and demonstrated by the preserved remains of *Plesiadapis* and *Palaechthon,* primates are basically primitive and generalized mammals in their bones. It was upon this primitive frame that subsequent primates became elaborated.

If the ancestral primate became different from other mammals in its teeth — those structures that not only dictate an animal's diet but procure and process it — then the next major leap of evolutionary inventiveness concentrated on the details of how the animal got about, visually and manually. This is where grasping hands and feet, the possession of a flattened nail at least on the big toe, and the development of enlarged orbits and a postorbital bar come in.

These are the features that further separate the prosimians and the anthropoids from the rest of mammals. These are the features that would have distinguished the common ancestor of prosimians and anthropoids from the first primate. And it is because of this ancestor that we, and all other living primates, are more visually than olfactorily oriented and are arboreally acrobatic. There would be no trapeze artists if this were not the case.

After the common prosimian–anthropoid ancestor, then what? If prosimians and anthropoids are a real evolutionary group, united because of common ancestry, are prosimians and anthropoids themselves natural groups?

The reality of Prosimii has not been easy to defend. It might be convenient to call prosimians those primates that one thinks are the most primitive, but in terms of tangible, definable features, Prosimii has traditionally been a wastebasket for those primates that are not anthropoids. With the tarsier supposedly representing a missing link between the "lower" and "higher" primates, a good deal of debate has been focused on just how close this pivotal primate is to one or the other of the groups between which it is sandwiched.

I have recently come to the conclusion that Prosimii, as a group, is real. All lemurs, lorises, bushbabies, and mouse lemurs have a spiky, clawlike "nail" on the second digit of the foot. The tarsier has one of these so-called grooming claws on its second pedal digit as well as on the third digit of its foot. It might seem like a pretty slim piece of evidence for proclaiming to have demonstrated the unity of Prosimii, but, as I have tried to emphasize, one cannot predict just how much "evidence" of relatedness there will be and just how it will be manifested. It might just be in this instance that there will not be much beyond the development of a grooming claw that will as easily indicate the relatedness of Prosimii.

There is another feature that I think unites Prosimii, but its defense involves a more complicated discussion of the evolutionary relationships among all potential living and fossil prosimians. Without getting too far off on a tangent, I can at least introduce you to this specialization.

With the exception of the aye-aye, with its large, continually growing anterior teeth, all lemurs, all lorises, all bushbabies, and all mouse lemurs have a set of modified teeth at the front of the lower jaw. In these primates, the lower incisors and the canines next to them are elongate, slender, and tilted forward, sometimes quite considerably. This dental "unit" has been called a tooth comb, in part because it looks like a comb with its narrow teeth packed ever so tightly together, and in part because all of these primates use this set of teeth in grooming their fur. Among all other living mammals, there are none that have anything specifically like the tooth comb of these prosimians. Therefore, we can suggest that the common ancestor of this aggregate of prosimians was distinguished by its development of a tooth comb. Because the affinities of the aye-aye lie well within the lemur group, as far as Ian Tattersall and I can figure out, we are forced to conclude that, in the course of the evolution of this odd primate, the tooth comb was lost — replaced by rodentlike anterior teeth.

On the face of it, it would appear that the tarsier might be a problem. Since it does not have a tooth comb, it could be just the sister of all other prosimians, to whom it is related because of sharing the novelty of a grooming claw. However, in a review of the whole issue of the place of *Tarsius* in primate evolution, I discovered an unexpected amount of evidence indicating that the tarsier was actually related to the lorises, bushbabies, and mouse lemurs. The supposed intermediate between the "lower" and "higher" primates more probably had its evolutionary ties within the group of prosimians. In this context, an earlier paper I had written on theoretical aspects of evolutionary tooth loss in mammals, and *Tarsius* in particular, made some extra sense. I had there come to the conclusion that the tarsier had lost its complement of lower incisors and canines. So it could very well be the case, then, that the reason *Tarsius* does not have a tooth comb, as virtually all other prosimians do, is because the tarsier no longer has the teeth necessary to make a tooth comb.

There is some fossil evidence that could be brought to bear on the issue of tooth combs. But with only the aye-aye and the tarsier lacking this dental specialization — lacking it not because they have the teeth and the teeth are unspecialized, but because they just don't have all the necessary teeth — one can reason that the common ancestor of at least all living prosimians had a tooth comb in addition to its specialized pedal grooming claw. Within Prosimii, the lemurs, lorises, bushbabies, mouse lemurs, and tarsiers diversified in many exotic directions.

The ancestral anthropoid — the common ancestor of New World monkeys, Old World monkeys, and hominoids — built evolutionarily upon the emphasis on a visual–manual complex. In anthropoids, the dependence on olfaction, smell, became further diminished, as is indicated, for example, by the fact that the nasal capsule of these primates is absolutely as well as relatively smaller than it is in prosimians. Vision became even more the dominant sense in anthropoids, as is noted, for instance, in their orbits being more consistently directed forward. Postorbital closure is one reflection of this change. Presumably, the reason that only monkeys and hominoids are distinctive in these characteristics is because these primates had a common ancestor to whom these attributes were unique.

In its teeth, the ancestral anthropoid would have been distinguished from the common prosimian–anthropoid ancestor in a number of features. In the course of the evolution of the ancestral anthropoid there would have been a further lowering of cusp height and a reduction in number of teeth, from four to three premolars in each quadrant of the mouth. There would have been the loss of the anteriormost cusp on the lower molars, but a new cusp would have developed as an addition to the posterior part of the upper molars.

The lowering of cusp height probably marks some change in diet, perhaps an increased consumption of fruit and leaves. The loss of a tooth within each row is probably associated with the facial shortening that occurred as a result of the reduction of the snout. I don't know why a cusp would be lost in the lower molars and one added to the uppers, but the effect of these alterations was to further square up the molar teeth. The loss of the forwardly projecting anterior lower molar cusp gave a straighter face to the front of the lower molar. You can see on your own lower molars just how squared up the front part of each tooth is. The addition of a fourth cusp to the upper molar on the posterior and tongue side of the tooth created a corner and thus a more rectangular tooth. Your first upper molar is the best example in your mouth of this configuration. You should be able to feel this extra upper molar cusp with the tip of your tongue.

In the course of fetal development, all mammals come to have separate right and left mandibles. Most mammals are born with separate right and left mandibles held together at the midline of the jaw by cartilage. Only in old age might there be some ossification, or bony deposition, in this midline cartilage which causes immobility of the two mandibular halves. Otherwise, the right and left mandibles are unfused at this midline symphysis and can move a bit independently of each other. Anthropoid primates, however, either are born with a unified mandible or fuse the mandibular symphysis early in life.

All mammals develop and are born with separate and distinct right and left frontal bones, which together make up the forehead region of the skull. These two ontogenetically individual bones may eventually fuse along a midline suture — the frontal suture — but such fusion is typically late in life, and evidence of this suture persists well into old age. In anthropoid primates, the frontal suture (called the metopic suture in humans) fuses early in life. Usually, all traces of it are obliterated in juvenile monkeys and hominoids.

Most female mammals have a uterus composed of two chambers that are confluent at the base, or cervical region. Each chamber is associated wih an ovary–fallopian tube system. Each chamber looks somewhat like a horn, and thus the common mammalian uterine configuraton is referred to as bicornuate.

In some mammals the two uterine cornua fuse to form a single-chambered, or simplex, uterus. A longitudinal furrow along the wall of the single uterine chamber attests to its derivation from what had ontogenetically been two separate units. Among the few mammals that develop a simplex uterus are anteaters and anthropoids.

In mammals there are three major configurations of the placenta, the connection between the developing fetus and the mother. These configurations reflect the varying degrees to which fetal membranes invade the uterine wall and associate with the mother's blood system. The most invasive connection, called hemochorial placentation, is found in only a handful of mammals. Various rodents, anteaters, the tarsier, and anthropoids develop hemochorial placentation. But each kind of mammal achieves its brand of hemochorial placentation through a somewhat different developmental path, so that we can identify different forms of this apparently similar end product.

Anthropoids are distinguished by having a type of hemochorial placentation that is especially characterized by the concentration of blood vessels into two disklike structures. Anthropoids are also distinctive in that the development of the membranes of the amniotic sac that surrounds the growing fetus proceeds by an infolding of cells. This unique combination of fetal membrane and placenta in anthropoids could be correlated with the development of a simplex uterus.

If at this point we step back for a moment to assess where humans stand evolutionarily, we find that humans possess features that would have specifically characterized the ancestral primate, others that would have been unique to the common prosimian–anthropoid ancestor, and still others that would have emerged with the common ancestor of monkeys and hominoids. Thus, features such as a short face, a postorbital bar and postorbital closure, grasping hands and the anatomy of grasping feet, nails instead of claws, a reduced number of premolars, lowered cusp height and rather rectangular molars, early fusion of frontal sutures and mandibular symphyses, the development of a simplex uterus, and discoidal hemochorial placentation, while certainly part of our makeup and important for our survival, are features that we possess in spite of our being human or even hominoid primates. We inherited these features from a history of successive ancestors.

And we are still not at our evolutionary doorstep. We must pass through yet another set of ancestors before we reach anything that we might isolate and identify as being specifically hominoid, much less human or ape.

New World monkeys, Old World monkeys, and hominoids are the anthropoid primates. Among other things, they share uniquenesses of their reproductive physiology, they develop postorbital closure, they have nails on their digits, they have short snouts, and so on. But within Anthropoidea, are the monkeys, New and Old World, a group, and are hominoids in turn related to this large congregation? Although they might all be called monkeys, Old World monkeys and New World monkeys are distinctly separate groups of anthropoids. It does appear, however, that hominoids and Old World monkeys are the more closely related among the anthropoid primates.

Old World monkeys and hominoids have long been distinguished by systematists from New World monkeys by the orientation of their nostrils. In fact, the taxonomic names given to extant New World

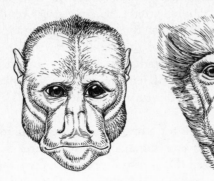

FIGURE 5F   A New World monkey (a red uakari, *Cacajao rubicundus*) *(left)*
and an Old World monkey (a Japanese macaque, *Macaca fuscata*) *(right)* to
illustrate the different configurations of the "platyrrhine" *(left)* and "catarrhine"
(right) nose. In the platyrrhine condition, the nostrils are broadly separated and
are oriented outward. In the catarrhine nose, characteristic of all living Old World
monkeys and hominoids, the nasal septum is narrow and the close nostrils are
more forwardly positioned.

monkeys and the group consisting of extant Old World monkeys and
hominoids reflect these nasal differences.

The New World monkeys have traditionally been placed in the
infraorder Platyrrhini, which means broad-nosed. This refers to the
fact that the nostrils of a New World monkey are separated from
each other by a rather broad, fleshy pillar (septum), as, for example,
one sees in a mouse, a ferret, or a lemur.

In contrast, Old World monkeys and hominoids are catarrhine —
narrow-nosed — primates. Catarrhinism results from a reduction of
the nasal cartilages that otherwise expand the nose of a platyrrhine
animal, causing a much more reduced and thinner nasal septum between the nostrils. Since catarrhinism is so rare among primates, as
well as mammals in general, it would appear that "narrow-nosedness"
is indeed a feature that would delineate Old World monkeys and
hominoids as a group. It is doubtful that we will ever discover a
fossilized nose, much less the nose of the ancestral catarrhine primate.
However, I think it is reasonable to conclude that the last common
ancestor of all extant catarrhines (which may or may not actually
subsume all potential fossil catarrhines) was catarrhine in its nose.

There are some other features that probably would have particularly characterized the catarrhine ancestor. In comparison with the ancestral anthropoids, the common ancestor of Old World monkeys and hominoids would have had a somewhat more expanded brain, and there would have been an increase in the size of its orbits accompanied by virtually complete postorbital closure. The region of the base of the skull that houses the tympanic membrane (ear drum), the tympanic ring across which this membrane is stretched, and the inner ear bones would have been denser, flatter, and less vacuous, and from the typmanic ring a stout, bony tube that communicated with the fleshy ear would have emanated laterally.

The face of this catarrhine ancestor would have been even more foreshortened, and the number of premolars would have been reduced from three to two in each quadrant of the jaw. The upper molars, especially the first and the second molars, would have been more square than rectangular, because of expansion of that posterior fourth cusp. Cusp height might also have been a bit lower, but canine teeth, especially in the male, would most likely have been large and powerful. Indeed, males were probably noticeably larger physically than females, but by how much is very difficult to reconstruct.

The postcranial skeleton of the ancestral catarrhine would have been modified somewhat, especially in the upper arm, which would have been slenderer, with the crests and ridges for muscle attachment de-emphasized.

Within Catarrhini, Old World monkeys are distinguished from hominoids by the arrangement of cusps along the premolars and molars into crosswise ridges, or lophs.

So far, we have followed the development of all sorts of aspects of the skull, teeth, postcranial skeleton, and even soft-tissue features, which are obviously parts of a functioning ape or human but have nothing to do specifically with being a hominoid, a human, or an ape. But we now can distinguish a hominoid primate, especially if we have the entire skeleton. The gibbons and siamangs of southeast Asia, the chimpanzees and gorillas of sub-Saharan Africa, the orang-utans of Borneo and Sumatra, and all humans are unique among primates in the shape of their trunks, in the chest as well as in the pelvic region, and in modifications of their arm and shoulder girdle.

In typical quadrupedal animals, the pelvis is a very narrow structure. You can feel this the next time you pet a dog or a cat. But a quick look at yourself in a mirror, or perhaps a peek at a passerby, illustrates immediately that a human's pelvis is deep from front to back and relatively broad from side to side.

The major component of the pelvis that concerns us here, and the part whose outline you can easily palpate below your waist, is a bladelike structure called the ilium. In quadrupedal mammals, the ilium is quite narrow and elongate. In hominoids in general, the iliac blade is noticeably shorter and deeper. In the four large hominoids, these dimensions are further exaggerated, and in humans, this blade is relatively the shortest and the deepest. Since all hominoids carry their trunks at least to some extent more vertically than do monkeys or lemurs, the added breadth of their pelvic region may function in "cupping" the abdominal viscera.

The pelvic region is not the only part of a hominoid to have become broad. The upper body of a hominoid has also been modified. In most mammals, the chest, or thoracic region, is narrow from side to side and deep from its top (vertebral border) to the sternum (breastbone). A horse, for example, is quite deep-chested, but its shoulder blades, or scapulae, are not broadly separated from each other. The thoracic region of a cat, a lemur, a loris, or any New or Old World monkey is also narrow from side to side and deep from vertebral column to sternum.

The thoracic region of a hominoid has a markedly different shape. Instead of being narrow and deep, the chest of a hominoid is broad from side to side and narrow from spine to sternum. A reflection of this overall configuration is the shape of a hominoid's rib, which is quite rounded from its vertebral articulation to its sternal attachment. Contrast this with the rib of a typical quadruped. If you think about a sparerib or a rib lamb chop, you might recall that these ribs, although having a slight curvature to them, are relatively straight.

A consequence of compressing the thoracic region of a hominoid from front to back is that the position of the hominoid scapulae must also be altered. In typical narrow-chested, quadrupedal animals, the shoulder blades are positioned on the sides of the rib cage. The broad, flattened hominoid chest forces the scapulae toward the back, close to the vertebral column. Since the head of the humerus (upper arm bone) articulates in a shallow socket at the end of the scapula, the position of the arm is also shifted, becoming emplaced on the side of the body. The arms of a hominoid have become broadly separated.

But this is not all of it. The collarbone, or clavicle, which extends like a strut between the sternum and scapula, must also readjust to these anatomical alterations. And so there is something special about the hominoid clavicle. The possession of a clavicle may not, in and of itself, be significant within Primates, but the development of a long clavicle is. And hominoids have one.

The clavicle is connected to a fingerlike extension of the scapula that projects just above the articulation with the head of the humerus. You can feel this on yourself by tracing the spine of the scapula forward, toward and over the shoulder joint. The clavicle and this extended scapular process form the upper part of the socket for the shoulder joint. This scapular process is particularly elongate in hominoids.

All hominoids, large or small, are distinguished further in their arms. Their humerus is quite elongate and its shaft is relatively narrow. The elbow joint is a much more detailed unit, with more ridging of the lower end of the humerus and a longer projection of the upper end of the ulna, one of the lower arm bones. The ridged lower end of the humerus creates a tighter articulation with the upper ends of the radius and ulna of the lower arm. The better fit of the bones of the elbow joint is also due to the development of the more elongate extension of the ulna, which locks into the back of the lower end of the humerus. At the wrist, the lower end of the ulna is more removed from articulation than in monkeys, thus giving more freedom of rotation and movememt at this joint.

All in all, the arm and shoulder girdle of a hominoid are distinct. The broadly separated shoulders, with great freedom of movement at the shoulder joint, the configuration of the elbow, and the greater mobility of the wrist have been interpreted as indicative of a major locomotive shift in the emergence of the ancestral hominoid.

A shift to even greater arboreal acrobatics.

A shift to a below-the-branch arm-swinging, catapulting, brachiating mode of locomotion, in which the trunk is held vertically — just the right position from which to descend bipedally from the trees.

6

The Hominoids

BY THE TIME we get to considering the emergence of the hominoids, we have pieced together much of an entire organism. A hominoid's skeleton is basically of general mammalian design. A hominoid's dental pattern was established tens of millions of years ago and put into near final shape with the emergence of the narrow-nosed anthropoid, the ancestral catarrhine. In fact, this ancestral catarrhine, the common ancestor of Old World monkeys and hominoids, was in many ways a hominoid in need of but a few finishing touches.

Those touches, however, were profound.

With only a few modifications in the shape of the torso, a new type of animal emerged — one with the ability not only to climb but to swing through the trees. This type of free-hanging, swinging locomotion is called brachiation. And in real brachiation (as opposed to imitation brachiation, which a few New World monkeys are capable of), the body is fully suspended below the arms and the animal propels itself with a hand-over-hand motion.

All hominoids, large or small, are brachiators — at least in theory. The highly mobile shoulder joints allow for wide excursions of rotation. The wrist joint is modified to accommodate the twisting that accompanies the path of the body's swing. But we are not all equally proficient brachiators, nor do all hominoids actively brachiate. Gorillas, for instance, are predominantly terrestrial animals, even though their anatomy is indelibly stamped with some of the hallmarks of brachiation. In fact, of all the hominoids, the gorilla has the possibility of the freest movement during brachiation because the lower end of its ulna is the most distantly separated from the neighboring wrist bones. On the other end of this spectrum are the gibbons and siamangs, and they are the world's ace brachiators.

Gibbons and siamangs have traditionally been placed in their own family, Hylobatidae, within the superfamily Hominoidea. Gibbons are distributed rather widely throughout southeast Asia, from the mainland into the islands of Indonesia, where their territories often overlap those of orang-utans. Siamangs are confined primarily to the Malay peninsula, where their ranges can overlap with those of gibbons, but they are also native to the island of Sumatra, where they can be found in the same areas with gibbons and orang-utans.

Gibbons and siamangs have small faces and round, globular crania with large, pensive eyes. They are rather small animals, as is readily apparent, for example, by the fact that the skull of a gibbon can fit easily into the palm of your hand. A siamang's skull is not more than ten percent larger than a gibbon's.

When standing upright, a gibbon is barely a few feet tall. Its legs look about normal, being a bit longer than its trunk. But a gibbon's arms are another matter. The gibbon's arms are so long in proportion to its body that the animal's hands drag on the ground.

Gibbons and siamangs live in the highest reaches of the Asian tropical rain forest and subsist primarily on fruits and leaves. They live in pairs, one adult male and one adult female, bonded for life. They are the only truly monogamous species of primate. And males and females share responsibilities equally, from attending to their offspring to defending the borders of their territory. In many cases, the female is even more aggressive than the male in hooting and calling and charging a potential trespasser.

Sociobiologists argue that there is a correlation between mating systems and the degree of sex-related differences, or sexual dimorphism, characteristic of a species. When males are larger than females, there seems to be a tendency for polygyny, the association of one male with more than one female. When females are larger than males, there tends to be the reverse, polyandry. When there is no size difference, there is monogamy. Human males tend to be a bit larger than human females, sometimes up to ten percent larger. But among gibbon and siamang males and females, there are no significant size differences.

With true brachiation, as seen in the gibbon and siamang, the animal's trunk is held vertically and erect as a result of the animal's suspending itself from an overhanging support, and the head is balanced on this vertical support. This position — a vertical vertebral column with the skull balanced centrally atop — is descriptive of humans as well.

Given this seeming similarity between gibbons and siamangs and humans, some paleoanthropologists, as we saw in chapters 3 and 4, sought to derive human bipedalism from a gibbonlike brachiator. From the waist up, this potential progenitor of bipedalism was already prepared, argued the paleoanthropologists. All that one needed to figure out were the selective forces that would have caused such an animal to drop to the ground and land on its two feet.

But perhaps this was too simplistic a view. The similarities in having a vertical vertebral column on which the skull was centrally balanced could be evolutionarily insignificant. Perhaps those apes that are more closely related to humans than gibbons and siamangs would make better evolutionary models.

Regardless of the specific relationships among the four large extant hominoids, the hominoids historically considered by most systematists to be most primitive are the gibbons and siamangs. This general consensus has in part resulted from the perception of the gibbons and siamangs as being hominoid enough to be admitted into the fold, but not being as humanlike, or anthropomorphous, as the large apes.

This is the type of approach that Thomas Huxley used in his book *Man's Place in Nature* to argue that not only were humans related to a specific group of mammals (the primates) but that humans were related closely to a specific subgroup (the apes) of this group of mammals. Adolph Schultz, the Swiss primatologist who defended the unity of the great apes as a group to which humans were related, also relied on this form of phylogenetic reasoning. However, as I argued in chapter 4, this approach — and any approach that does not attempt to distinguish those similarities that are due to *recent* common ancestry from those inherited from more distant ancestors — cannot necessarily be expected to yield accurate conclusions of relatedness.

Nevertheless, even when the "rigorous" approach has not been applied, a conclusion sometimes emerges that stands up to attempts of rejection. There does appear to be good reason for thinking that gibbons and siamangs are the sister taxa of a group of large hominoids. This conclusion is the result of a demonstration that, within Hominoidea, the large hominoids themselves constitute a group — that there are evolutionary novelties that humans, orang-utans, chimpanzees, and gorillas share which they do not also share with gibbons and siamangs.

Sometimes it is not always possible to delineate features that herald the relatedness of taxa you take for granted as constituting a group. For example, Old World monkeys may be united easily on the basis of their development of bilophodonty. But that doesn't mean that those catarrhines that are left over necessarily have anything evolutionarily spectacular to recommend them as a real evolutionary group.

In this case, however, we were lucky. There do appear to be evolutionarily significant features supporting the unity of Hominoidea. As we began to see in chapter 5, features strikingly distinctive of this assemblage of catarrhine primates are departures from the more common mammalian design in the shape of the rib cage, the position of the scapula, the length of the clavicle, the configuration of the humerus, the bony contributions to the elbow joint, the ulnar articulation in the wrist, and the shape of the ilial blades of the pelvis. But that was only the beginning. There are still a few more features that seem to attest to the evolutionary unity of the hominoids. And some of these attributes are only the beginning of a rather unusual series of specializations within Hominoidea — specializations or evolutionary novelties that will emerge as common to an unexpected twosome of hominoids.

The sacrum, a bony structure at the end of the vertebral column, separates the right and left halves of the pelvis. In adults, the sacrum appears to be a single bone, but it is really the end product of the fusion of ontogenetically separate vertebrae. Knowing this, one can easily see on an adult sacrum the traces of this early phase of vertebral distinctiveness.

Hominoids are distinguished among catarrhine primates in having a greater number of sacral vertebrae than Old World monkeys. On average, Old World monkeys have only three sacral segments. Gibbons and siamangs, however, generally have four or five sacral segments. The other hominoids have even more, either five or six.

Adolph Schultz portrayed this difference between Old World monkeys and hominoids in number of sacral segments as an advance upward of the pelvic region in the hominoid catarrhines. In hominoids, the pelvis "captures" more lower or lumbar vertebrae to create a longer sacrum. But this marching of the pelvic region up the vertebral column brings with it another alteration, one that affects the next set of vertebrae, the lumbar vertebrae, which sit atop the sacrum. The lumbar region of a hominoid's vertebral column becomes shorter. While Old World monkeys typically have seven lumbar vertebrae, hominoids have five or fewer of these vertebral elements. The lumbar vertebrae of hominoids are also much thicker and relatively stouter.

Increasing the length of the sacrum and shortening and thickening the lumbar region — resulting in a less flexible lower torso than monkeys have — is apparently correlated with a shift in the way hominoids carry themselves. More specifically, these skeletal changes would have been correlated with the way the ancestral hominoid carried itself.

There might also be a correlation between changes in the sacral and lumbar regions of the vertebral column and the fact that all extant hominoids lack caudal vertebrae. Caudal vertebrae are those elements that make up the "backbone" of an animal's tail. The only vestiges we and other hominoids have of our more distant relatives' tail is the persistence of a few tiny bits, called coccygeal vertebrae, that are tacked onto the end of the sacrum. In contrast to all other primates, the common hominoid ancestor would have been distinguished among catarrhines by its taillessness.

Adolph Schultz was very interested in differences among the hominoids, not just in the adult but throughout an individual's developmental life and the degenerative phases of old age. One of his studies focused on the state of development at birth of the bones of the arm and hand of catarrhine primates. And this, too, proves enlightening with regard to our understanding of the evolution of hominoids.

When we think about the workings of a skeleton, we have an image of bones fitting nicely together with the end of one bone having a shape that complements the shape of the end of another bone, with which it articulates. For example, the lower, or distal, end of your thigh bone, or femur, has two rounded surfaces that fit onto two slightly concave surfaces on the upper, or proximal, end of the tibia, while the head of the femur is a ball-shaped structure that fits into the pelvic socket of your hip. But these bone ends, called epiphyses, were not always attached to the bone shafts.

Ontogenetically, epiphyses are separate elements that have their own schedules of development and ossification. For example, the epiphysis of the articular head of the femur appears in humans before the sternal end of the clavicle begins to develop. In part because of these differential rates of development and ossification, epiphyses also fuse at different times to the different bones. The femoral head begins fusion at about fifteen years of age and is completely fused to the shaft of the femur by twenty or so years, whereas the sternal end of the clavicle begins to fuse to the body of the clavicle at about eighteen years of age and does not become solidly joined until at least the age of thirty. To make matters even more complicated, the proximal and distal epiphyses of the same bone do not necessarily have the same developmental schedules nor do they fuse at the same times to their shared bony shaft. For example, the distal end of the femur begins fusion to the shaft of this long bone at about sixteen years of age, a year or so after the proximal end begins its fusion, and completes its union with the femur at about twenty-three years, which is three or more years after the proximal end has fused completely. While they are unfused, epiphyses are separated from the shaft of the bone by a layer of cartilage. As long as the epiphyses remain unfused, the bone can grow in length. When fusion between epiphysis and bone shaft occurs, longitudinal bone growth ceases.

Since Schultz collected data only on the arm of catarrhine primates, this discussion is unfortunately confined to this region of the skeleton. But I do not think it would be greatly inaccurate to suppose that, if there are developmental differences among animals showing up in their arm bones, similar differences of timing would not also affect their leg bones.

Typical quadrupedal Old World monkeys, such as macaques, gelada baboons, and savanna baboons, are born with ossification having begun in all bony elements, including all the epiphyses, of their arms. Hominoids are different from Old World monkeys in the timing of the onset of ossification of the epiphyses of arm and hand bones. In all hominoids, ossification of the epiphyses that contribute to the elbow joint is delayed: at birth, the lower epiphysis of the humerus and the upper epiphysis of the radius have not yet begun to harden. Whether or not there is a correlation between this developmental delay and the morphology that will eventually materialize, it is nevertheless also the case that the distal end of the humerus of hominoids is distinct among primates in its shape. Hominoids possess a midline ridge on the distal end of the humerus which creates a tighter articulation of the elbow joint.

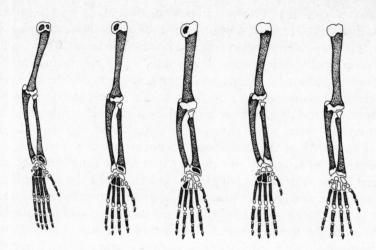

FIGURE 6A The arm of a gibbon, a chimpanzee, a gorilla, an orang-utan, and a human *(from left to right)* illustrating the states of ossification of epiphyses at birth (all drawn to the same length). Notice, for instance, that the four large hominoids delay ossification of one of the two centers of the proximal humerus epiphysis and that orang-utans and humans go even further in delaying both of these centers of ossification.

Hominoids also delay the onset of ossification of some of the carpal (wrist) bones. In newborn Old World monkeys, all seven of the carpal bones may be found to have begun to ossify. In neonatal hominoids, however, the rule is that few of the carpals have begun to ossify. The majority of mammals are born with ossification proceeding apace in virtually all components of the skeleton. We can think of this as development proceeding at a normal rate toward adulthood. A re-tardation of any of these developmental rates would result in an extension of the more juvenile aspects of the individual. Heterochrony refers to all of the possible ways in which rates of physical and sexual development and maturation can be altered and the ways in which they would affect each other. One of the results of having different rates of physical and sexual development is called paedomorphism, which is a process of "fetalization." One of the results of paedo-morphism is called neoteny.

In neoteny, juvenile growth rates and/or juvenile features are re-tained in the adult. Humans, who look so "childlike" in their small faces with big eyes and their large, rounded crania, are usually cited as an example of neoteny at work. One of the primary demonstrations of neoteny in humans is the extent to which ossification has been retarded in the newborn. A human's skeleton at birth differs from that of most mammals in the degree to which it remains cartilaginous.

Obviously, humans are not the only catarrhines to be distinguished by some noticeable delay in ossification of various elements of the arm and wrist. Although humans are more "neotenic" than most other hominoids, all hominoids are at least somewhat neotenic in delaying ossification in the distal humerus, proximal radius, and many of the carpals of the wrist. That suggests that some degree of neoteny existed in the ancestral hominoid — that neoteny is important in hominoid phylogeny.

The smaller hominoids, the gibbons and siamangs, remain similar to Old World monkeys in that the proximal epiphysis of their humerus has begun to ossify by birth. Gibbons and siamangs also remain similar to Old World monkeys in initiating ossification of the distal ulnar epiphysis before birth. The large hominoids are, however, different among catarrhines. At birth, ossification has not yet begun in one segment of the proximal humeral epiphysis, nor is ossification evident in the distal ulnar epiphysis.

The delay in ossification of the lower ulnar epiphysis is particularly interesting in light of further alterations in the wrist of the large hominoids. If all hominoids, large and small, are set apart from other primates in that there is some disruption of direct and constant contact between the distal end of the ulna and the neighboring carpals of the wrist, then the large hominoids are even more distinct in that the lower end of their ulna is even farther removed from direct articulation with the carpals.

Gibbons and siamangs remain similar to Old World monkeys in the initiation by birth of the ossification of the distal metacarpal epiphyses. Metacarpals are those bones that articulate between the carpals of the wrist and the upper bone (first phalanx) of each digit of the hand. There is a metacarpal for the thumb and one for each finger. The distal metacarpal epiphysis articulates with the upper end of the first phalanx of the thumb or finger. In newborn chimpanzees, gorillas, orang-utans, and humans, these distal metacarpal epiphyses have not yet begun to ossify.

By their delaying further epiphyseal ossification of the metacarpals, of the distal end of the ulna, and of a component of the upper humerus, and by their decreasing further the contact between the ulna and its neighboring carpals, we can begin to delineate a group within Hominoidea. The group is the large hominoids — the chimpanzee, the gorilla, the orang-utan, and humans. Their very size immediately distinguishes them among extant primates, and the delays in forelimb ossification and the decoupling of ulnar–carpal articulation further set them apart. This is only the tip of the iceberg. There are other aspects of their anatomy and especially of their reproductive physiology that are perhaps even more evolutionarily compelling.

All mammals produce and excrete various sex hormones, or steroids. There are two rather broad categories of sex hormones, androgens and estrogens. Although each sex maintains levels of each kind of steroid for at least part of its lifetime, androgen is the hormone generally associated with maleness while estrogen is typically identified with femaleness. Above-normal levels of estrogen in a male will produce enlarged breasts and a femalelike deposit of fat around the hips and thighs. Above-normal levels of androgen will produce excessive hairiness in a female.

Biochemical analysis of urinary excretions can detect steroids and delineate the various estrogens or estrogen products manufactured by an individual. In most female primates, the major urinary estrogen is a steroid called androsterone. Rhesus monkeys, for example, excrete high levels of androsterone. However, in the large hominoids, the major component of estrogen is a steroid substance called estriol. During the menstrual cycle, estriol is involved, for instance, in the induction of ovulation. During pregnancy, estriol affects the placenta, the network of membranes connecting fetus with mother. The large hominoids are distinguished from other primates by their higher levels of estriol excretion during the menstrual cycle and pregnancy. The monitoring of estriol levels during pregnancy gives us an indication of how the fetus is developing because there appears to be a correlation between maternal estriol levels and the secretion by the fetus's adrenal glands of another steroid, a variant of androsterone.

The adrenal gland is one of the organs that produce steroids. An animal has two adrenal glands, which sit on top of the kidneys like mushy caps. Fetal adrenal secretions seem to be correlated with brain growth. For example, in humans with reduced brains, there is a reduced fetal adrenal zone of excretion and the adrenal glands themselves are also notably reduced in size.

Bill Lasley, Nancy Czekala, Susan Shideler, and others at the San Diego Zoo have been studying these hormonal levels and have begun to arrive at some interesting conclusions. Differences between animals in their levels of estriol excretion during pregnancy are not just reflections of differences in the placenta. Rather, they are probably due to differences in fetal development.

Since the large hominoids are distinguished from other primates by their high levels of estriol excretion, we have additional reason to accept the evolutionary reality of a large hominoid group. And because there appears to be a correlation between maternal levels of estriol and fetal development, we might go further and suggest that the unity of this evolutionary group is also reflected in distinctions of fetal development. If we can correlate maternal estriol levels with fetal adrenal secretions, and if levels of fetal adrenal secretions are reflective of the extent of brain growth, cerebral distinctiveness might also be a feature that sets the large hominoids apart from other primates.

Not only do the large hominoids have large brains, their brains differ from those of other primates in having more distinct asymmetries to them, where features of the right and left cerebral hemispheres are unequal in size or extent of expression, as Norman Geschwind, Marjorie LeMay, and others at Harvard Medical School have shown. Humans have long been known to develop extreme cerebral asymmetries, which are supposed to be correlated with language and handedness.

If we turn to other parts of the body, we find that the ilial blades of a large hominoid's pelvis are even broader and more fanlike than in the gibbon and siamang, which, in turn, have broader ilia than Old World monkeys. The scapula (shoulder blade) of the large hominoids is also different from that of other catarrhines. One major difference is that a large hominoid's scapula is relatively shorter, from its area of articulation with the head of the humerus to the border that lies closest to the vertebral column. Essentially, the scapula of a large hominoid is a relatively much stouter and deeper blade than is the scapula of a monkey.

Teeth are not as blatantly revealing about the broader aspects of hominoid relationships, but when I reviewed the evolutionary relationships of all primates, fossil and extant, it struck me that there might be features that could be considered dentally hominoid. If my analysis was accurate, then the pattern of molar tooth cusps that would be primitive for hominoids would be the configuration we see in the gibbon and siamang.

The cusps of a gibbon's molar are basically puffy structures. Especially in the lower molars, the cusps are like tiny gumdrops pushed together so that their sides come into contact, flattening out a bit as they do. The tips of the cusps, or the crowns of the gumdrops, do not lose their separateness and integrity — they rise a bit above their bases.

The cusps of a large hominoid's molars are not puffy. They may be rather tall, as they are in a gorilla, or they may be quite low, as they are in orang-utans and humans. But these cusps appear more flattened than rounded, not only on their sides but especially on their occlusal (chewing) surfaces. The cusps of a large hominoid's molars are also more melded together at their bases, particularly in the chimpanzee, the orang-utan, and humans, than in gibbons and siamangs. These molar features would have distinguished the common ancestor of the large hominoids from the ancestor of all hominoids.

I think we can also conclude that the ancestor of the large hominoids had canine teeth that were significantly different from those of the other catarrhines. A minor stumbling block in the broader comparison is the fact that modern humans have very tiny and insignificant canines. But since all other extant catarrhines — Old World monkeys, gibbons and siamangs, the two African apes, and the orang-utan — and virtually all fossil catarrhines have or had large canines, a reasonable conclusion is that reduction in canine size would have been an isolated and special aspect of hominid evolution.

The canines of Old World monkeys and gibbons and siamangs are tall, pointed, stilettolike teeth. This is particularly true of the upper canines, which are usually much longer than their lower counterparts. One aspect of sexual dimorphism in catarrhines is canine size — the canines of the male tend to be noticeably larger than those of the female. A yawn from a threatening male baboon, exposing two inches or more of upper canine, is usually sufficient to thwart the challenge of a subordinate male seeking to take over the troop. A front line of adult male baboons all exposing their dental "artillery" can cause hunting lions or hyenas to back away and consider another source for their next meal.

The canines of a chimpanzee, a gorilla, and an orang-utan are pointed, and the upper canine is also longer than the lower canine, but these are not stilettolike. They are much stouter and bulkier, and rather broad at their bases. These canines are also relatively shorter than they are in an Old World monkey or a gibbon or siamang, even though the tips of these teeth do extend noticeably beyond the occlusal surfaces of the neighboring teeth. Thus, while having large canines is not unique to any of the three large apes, these hominoids have large, stout canines that are broad at the base, and presumably the ancestor of the large hominoids did too.

The large hominoids — including humans — are also distinguished among catarrhine primates in bony aspects of their nasal region. First of all, the shape of the bony margin or edge of a large hominoid's nasal aperture is distinctly different from that of other primates.

You can palpate the margin of your own nasal aperture. Its limits are where the fleshy nose extends from its sides and where the supporting cartilage is anchored above. As you can feel, the bone is narrower on top, where the cartilage articulates, and much broader at its base, behind the flare of your nostrils. The nasal aperture is thus essentially trapezoidal in shape, resting on the broader of its two parallel sides.

Only the large hominoids have this trapezoidal configuration of the nasal aperture. In all other primates — from the gibbons and siamangs, to the Old World monkeys, New World monkeys, and all prosimians — the aperture is ovoid or elliptical in outline.

There is another aspect of the nasal region in which the large hominoids are distinctive. It has to do with the configuration of the floor of the nasal cavity.

Think of cutting out the nose of a pumpkin for Halloween. Cut a trapezoidal nasal aperture, narrow on top and broad below, with sloping sides. If you were really accurate with the knife, and if the blade was so sharp that it left no ragged edges, the floor of the nasal cavity just inside the margin of the nasal aperture would be flat.

This would characterize the configuration most commonly found in the orang-utan. The floor of the nasal cavity of the orang-utan is typically quite flat. In the other large hominoids, the front edge of the lower margin of the nasal aperture may be thickened, so that if you were a miniaturized Alice walking through such a nasal aperture, you would have to step down a bit to get to the floor of the nasal cavity. The gorilla seems to have the most consistently thickened lower margin of the nasal cavity of the large hominoids and thus the most accentuated "stepping down" to its nasal cavity floor. A slight "stepping down" is sometimes observed in orang-utans, but it is restricted primarily to the skulls of juvenile individuals. In all other primates, from individual to individual and species to species, the lower margin of the nasal aperture is consistently very thick, causing a more noticeably "stepped-down" subnasal plane than in the large hominoids.

The upper jaw of a mammal is composed of two bones. The larger of the bones is the maxilla, which contains most of the teeth. The second bone is confined to the very front of the upper jaw, extending back not much farther than the nasal region. This bone, called the premaxilla, typically contains the incisor teeth. The lower margin of the nasal aperture is formed by the uppermost extent of the premaxilla, and it is this piece of bone that, when thickened, creates the "stepped-down" configuration of the nasal cavity floor.

If you were to look for the edges of contact, or sutures, that separate the premaxilla and the maxilla, you would find traces running along the face, between the nasal region and the edge of bone in which the teeth are anchored, as well as across the palate between the right and left canines. In most mammals, the course of the palatal portion of the premaxillary–maxillary suture is interrupted by two fairly large holes, one on each side of the longitudinal midline suture of the palate. If you were to hold up a skull of a deer or a dog, you could peer through these holes, from the palatal side, right into the anterior portion of the nasal cavity. These holes are like windows in the palate between the oral cavity and the nasal cavity. The anatomical name for such a hole is "fenestra," which actually means window. Most mammals have a pair of relatively large anterior palatine fenestrae. In the living animal, one does not see two large holes in the roof of the mouth. At most, the experienced biologist would note two tiny perforations in the flesh of the palate. Membrane and tissue are stretched across the fenestrae.

One day, after hours of staring at a table full of upturned skulls of gibbons, chimpanzees, gorillas, orang-utans, humans, and a representative assortment of Old and New World monkeys and prosimians, while trying to understand the changes that happen to the bone of the jaw while teeth are erupting, I found my mind wandering. I began looking at nuances in the shapes of anterior palatine fenestrae. Palatine fenestrae are usually fairly large, but they can vary from being teardrop shaped to being nearly oval or to looking like a quarter or third moon. There they were in prosimians, monkeys, and gibbons, but they were not present in the large hominoids.

The large hominoids do not develop large anterior palatine fenestrae. I had never heard of that before. In fact, I had never read anywhere anything peculiar about this region in any animal except *Homo sapiens*.

But there it was.

When I went to the literature to double-check my suspicions that this was a new discovery, I found no reference to this uniqueness of the large hominoids — even though, as long ago as the late nineteenth century, this condition had been illustrated accurately in figures of apes that accompanied published texts.

Not all large hominoids have the configuration characteristic of *Homo sapiens*. What all hominoids *do* have in common is what I have come to refer to as the "shutting down" of what we see in the palates of other animals as two fenestrae between the oral and nasal cavities.

Take two anterior palatine fenestrae, a right and a left. Shrink them in diameter and thicken the bone of the palate a bit. Now it looks as if two narrow canals have perforated the palate. Push these two canals together, toward the midline of the palate, until they abut each other but are kept apart by a thin wall of bone. This is how the palate of a chimpanzee or especially of a gorilla appears.

In the gorilla specifically, these canals are long and continue anteriorly along the palate as parallel grooves separated by a thin, bony septum. In the chimpanzee, the two canals are shallower and their distinctiveness not as marked, and in rare instances they may even become confluent.

Now push those canals together, so that a single foramen — aperture — perforates the palate. This is what we find in humans, the so-called single incisive foramen.

If you continue to push from the outside of what had been two canals, the single incisive foramen becomes much narrower and slit-like. This is what one sees in the orang-utan.

All large hominoids have "shut down" the large anterior palatine fenestrae. The African apes, the chimpanzee and the gorilla, typically only go so far as to bring the resultant canals into intimate association. Humans and orang-utans, however, have developed a palatal confluence between these two canals. And orang-utans go beyond

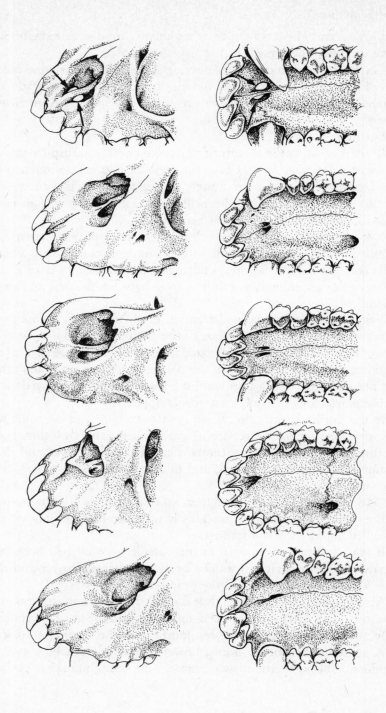

humans in that this single incisive foramen is compressed into a thin, slitlike channel.

What does this mean?

This type of evolutionary question — although one of the most common and frequently asked, and undoubtedly one of the most interesting type of questions — is also one of the most difficult and speculative.

I don't know what it means, at least in the sense of function. And I doubt that anyone else really does either. We can only guess — intelligently, one hopes.

Since catarrhine primates do not develop functional vomeronasal organs — that is, organs within the nasal cavity close to the anterior palatine fenestrae that intermingle taste and smell — there is the question of what, if anything, takes the place of them. Old World monkeys and gibbons still develop the large anterior palatine fenestrae through which right-to-right and left-to-left nasopalatine and greater palatine nerves must course conjointly. Without a viable vomeronasal organ, it would seem that these sensory loops would be lacking in sensitivity. But if you brought the rights into proximity with the lefts, and then even allowed for the possible contact between the right and left nerve loops, there might be a novel sensory compensation.

This seems to be worth considering. But one does not need to know the real function of an anatomical configuration to consider its implications for reflecting relatedness of organisms. We are interested

◄ F I G U R E 6 B   Views into the nasal cavity *(left-hand column)* and of the oral side of the palate *(right-hand column)* of *(from top to bottom)* the skull of a gibbon, a chimpanzee, a gorilla, a human, and an orang-utan (all drawn to the same size). The gibbon retains the common and primitive mammalian configuration of having two openings — the anterior palatine fenestrae *(arrows)* — that communicate between the oral and nasal cavities. The large hominoids have "shut down" these openings into canals. In the chimpanzee and gorilla, the two openings in the floor of the nasal cavity typically persist as two canals that open on the oral side of the palate; in the gorilla, these two canals are longer than in the chimpanzee and burrow somewhat forward along the palate. In humans and orang-utans, the two canals emanating from the floor of the nasal cavity — which are narrower than in the African apes — merge to form a single canal that opens on the oral side of the palate. In humans and other hominids, this single opening — the incisive foramen — is rather large and vacuous. In the orang-utan, as well as in *Sivapithecus,* this single incisive foramen is long and slitlike.

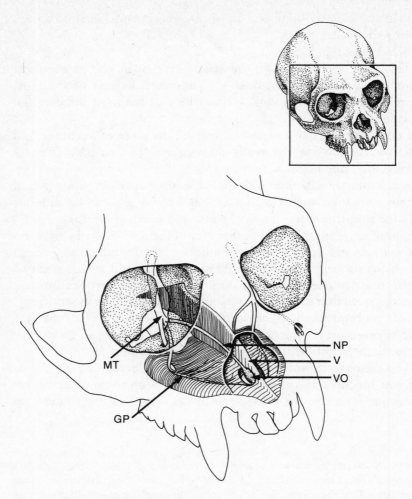

FIGURE 6C Breakaway view of some of the internal facial anatomy of a tamarin *(Leontopithecus),* a New World monkey, to illustrate the typical primate and mammalian disposition of these structures. Inside the nasal cavity, a bony plate, the vomer (V), separates this chamber into two components. Along each side of the vomer courses a branch of the maxillary nerve (MT), a division of the trigeminal nerve. The branch is called the nasopalatine nerve (NP) and is associated with a sensory organ, the vomeronasal organ (VO), situated in or near the palatine fenestra. Another branch of the maxillary nerve, the greater palatine nerve (GP), courses along the oral side of the palate and sends a few tiny branches into the anterior palatine fenestra which communicate with fine branches of the nasopalatine nerve.

in identifying homologous structures, which is difficult enough. And it is only those similarities resulting from common ancestry — those homologous similarities — that count.

One check on the homology of hard-tissue features, such as foramina, is the identity of the nerves or arteries that course through them. Because the same nerves — the nasopalatine and greater palatine nerves — course to the anterior palatine fenestrae as to a single incisive foramen, it is reasonable to conclude that the single foramen is derived somehow from the two fenestrae. This conclusion is also indicated by the persistence of two tiny canals in the nasal cavity, one on either side of the vomer (the bony plate that bisects the nasal cavity), that then become confluent as a single incisive foramen.

These conclusions gain further support from another approach to interpreting the phylogenetic significance of the different character states. Since most mammals possess two anterior palatine fenestrae, it would most definitely appear to be the case that this is the primitive condition for mammals, and that other configurations are more specialized, more derived. Deviations from the double-fenestraed condition are the evolutionary novelties.

"Shutting down" the palatine fenestrae would be unique among mammals. I am here referring to any configuration which cuts off broadly unrestricted communication between the oral and nasal cavities through two large palatine fenestrae. Among primates, this would describe the large hominoids, regardless of the finer details of difference that might exist among the four genera. As such, this general condition of "shutting down" the palatine fenestrae can be interpreted as being derived (relative to the common mammalian pattern) and thus reflective of the evolutionary unity of the large hominoids.

How can this suggestion be tested? A theory of homology can be tested by another theory of homology. The more robust — that is, corroborated — a set of theories of homology is, the more robust is the resultant theory of relatedness.

The theory that the large hominoids are closely related, that they shared a common ancestor not shared with any other extant primate, is quite robust. From further delays in forelimb ossification, to alterations of the scapula and ilium and complete removal of the distal end of the ulna from direct articulation with the carpals of the wrist, to the emphasis on estriol as the major estrogen and apparent changes in fetal development and brain growth, each suggested homology corroborates each other suggestion of homology. If one can conclude from this kind of data that organisms are closely related, there would seem to be little doubt that the large hominoids constitute a real evolutionary group.

But what about the details of how the four large hominoids differ in their atypically fenestraed states? They may all be characterized by having "shut down" the anterior palatine fenestrae, but some of them have done different things on top of this derived configuration.

Within the general context of having "shut down" the anterior palatine fenestrae to canallike passages between the oral and nasal cavities, there is an even more derived state. It is the development of a single incisive foramen palatally, as seen in *Homo sapiens* and long recognized as special. But there is an even more derived, compressed version of a single incisive foramen, and it is found in the orang-utan.

These are some of the thoughts that began to materialize as I tried to comprehend the significance of what I had come upon while studying that array of upturned skulls. The logical conclusion, given all the backup from nerves and comparative anatomy, was that the single incisive foramen in humans and the single incisive foramen in the orang-utan are homologous structures, derived from the "shut-down" condition that would have characterized the common ancestor of all the large hominoids. Humans and orang-utans had single incisive foramina because they had inherited this condition from an ancestor that they shared, but which they did not share with any other extant primate, the chimpanzee and the gorilla included.

Now here was a theory of homology — that the single incisive foramen in humans was derived from the same genetic and developmental stuff as the single incisive foramen in orang-utans. As with all theories of homology, both good and bad, this theory of homology generated a theory of relatedness, which suggested that, among the large hominoids, humans and orang-utans were more closely related to each other than either was to the chimpanzee, the gorilla, or the common ancestor of chimpanzees and gorillas.

During the early 1980s, orang-utans were elevated to a peak of attention because of certain fossil discoveries. Those discoveries *should* have resulted in a better and more detailed understanding of hominid evolution, but that was hardly what happened.

It was largely through the efforts of Elwyn Simons that the Miocene fossil hominoid *Ramapithecus* was "rediscovered" in the 1960s as a potential early hominid, perhaps even ancestral to *Australopithecus*. Over the years, few of the many features that Simons had used to argue for the hominid status of *Ramapithecus* survived the battering of close scrutiny. But this did not really matter, because it is not the *number* of features that is important, it is their evolutionary *quality*. The characteristics that did survive decades of heated debate were features of the cheek teeth, the premolars and molars.

Ramapithecus could be distinguished from the African apes and from virtually all fossil apes by its possession of low-cusped cheek teeth and a thick layer of enamel on the molars. Since "proper" hominids — *Homo* and *Australopithecus* — were also distinguished from the African apes by these dental attributes, most paleoanthropologists came to accept the notion that *Ramapithecus* was closely related to *Homo* and *Australopithecus*.

In the late 1970s, David Pilbeam, who had been collecting and analyzing Miocene fossil hominoids from the Siwalik hills of Pakistan, expanded the notion of a *Ramapithecus* group. The fossils *Sivapithecus* and *Gigantopithecus* had also been found to be "hominid" in their low-cusped cheek teeth and in their thick-enameled molars. Pilbeam suggested that there was a clade, or evolutionary group, of ramapithecids, which was somehow related to the later hominids.

After all the years of debate over which features of jaws and teeth really were diagnostically hominid, there was almost a sense of euphoria that came with settling on thick molar enamel and low-cusped cheek teeth as being the common denominator of our own — at last! — evolutionary group. Almost every time you turned around, someone had taken a section out of a tooth, or even a tooth out of a rather complete specimen, to take back to the lab, slice, polish, and then calculate how thick or thin the enamel was.

It turned out that the chimpanzee and the gorilla, like virtually all other primates, had thin-enameled molars. One could easily see that the cusps of a chimpanzee's and especially a gorilla's premolars and molars were also taller than in a ramapithecid or a hominid. But of course the cheek tooth features we see in ramapithecids and hominids could have evolved from the primitive primate condition retained in the African apes.

When the orang-utan was studied, it was found to have thick enamel covering its molars. And anyone could see that an orang-utan's premolars and molars are quite flat and low-cusped.

How could this be — especially if these dental features are supposed to be evolutionarily revealing about the finer details of relatedness among only a few of the hominoids, the ramapithecids and the hominids?

Given the prevalent view of who was related to whom among the large hominoids, the answer was simple. Those "hominid" features that were also found in the orang-utan had to have evolved independently and in parallel both in the orang-utan and in the common ancestor of ramapithecids and hominids because these features were interpreted in the context of a theory that maintained that the African apes — the chimpanzee and the gorilla — were the most closely related to humans among the hominoids.

In principle, a theory of relatedness is supposed to be based on theories of homology. If you choose one theory of relatedness over another, then you have effectively tested, or at least judged, one set of homologies against another and have decided that only one of these sets of similar features is actually the set of homologous features.

The suggestion that low-cusped cheek teeth and thick molar enamel in the orang-utan and in the ramapithecid–hominid group are homologous features results in the suggestion that these hominoids are related. But as far as I can tell, this suggestion was never entertained. Another theory of relatedness, which one assumed must have been based on another set of theories of homologies, took precedence. You would think that these theories of homology, indicating the relatedness of humans and the African apes, must have survived a history of intense scrutiny in order to have emerged as highly corroborated. Surely, the robustness of the theory of human–African ape relatedness must be virtually unquestionable for there to have been doubt thrown on the relationship of ramapithecids to "proper" hominids, and even on the very dental uniquenesses that would otherwise appear to be diagnostic of being hominid.

The year 1980 witnessed the first published description of part of a facial skeleton attributable to the fossil *Sivapithecus*. This potential ancestral hominid had those special facial details that were otherwise particular to only the orang-utan among extant primates. Since *Ramapithecus* and *Gigantopithecus* seemed to be intimately associated with *Sivapithecus,* the whole lot of ramapithecids was removed from near-hominid status and allocated to a position near the ancestry of the orang-utan. Along with this drastic maneuver, most paleoanthropologists came to the conclusion that low-cusped cheek teeth and thick molar enamel were no longer, and could no longer be, considered reflective of the relatedness of any specific subgroup among the large hominoids. Low-cusped cheek teeth and thick-enameled molars were seen as primitive retentions, or parallel evolutionary acquisitions, among those large hominoids with these features.

But it seemed to me that the cheek tooth data could provide a possible test of the homology of the single incisive foramen in humans and the orang-utan. Did all hominoids with low-cusped cheek teeth and thick molar enamel also possess a single incisive foramen?

To answer that question, I had to survey the available palatal regions of all species of *Homo* and *Australopithecus* as well as the ramapithecids, for these were the hominoids with "hominid" dental features. It would also be necessary to double-check the condition in thin-enameled hominoids. I scrutinized all the illustrations, descriptions, and casts of ramapithecids and hominids I had not had the opportunity to study in the original. I also surveyed all potential fossil apes. A pattern of cheek tooth and palatal morphology began to emerge.

The known palates of *Ramapithecus* and *Sivapithecus* preserve a single incisive foramen. In fact, one of the features that Peter Andrews and David Pilbeam had specifically singled out as uniting *Sivapithecus* with the orang-utan was the possession by both the fossil and the living animal of a long, slitlike single incisive foramen. Upper jaws of *Gigantopithecus* have not yet been discovered. I can only predict that this hominoid would have possessed a single incisive foramen, not double anterior palatine fenestrae.

All "proper" hominids have a single incisive foramen. It is large in all, just as in modern *Homo sapiens*. In the palates of Neanderthals, *Homo erectus,* whatever *Homo habilis* is, the various *Australopithecus*, including the hyper-robust Zinjanthropus type, and even in the preserved palates of Lucy, Johanson and White's *Australopithecus afarensis*, there is a single incisive foramen. It seemed that, among hominoids, those with "hominid" dental characteristics also had a single incisive foramen.

Of course, when a picture is coming together, there is always the anticipation of finding the exception to the rule. I go to museums now turning over the skulls of any and all hominoids, fossil and extant — like the old shell game — to see what lies underneath, in the palate.

But so far my original observations are fairly well confirmed. Recently I have had opportunities to study more of the original fossil hominid material, and they, too, bear the stamp of a single incisive foramen.

I roughed out a tentative scheme of hominoid evolution. It went something like this.

First, there were features of the postcranial skeleton in particular that indicated the unity of hominoids as an evolutionary group. Second, there were also features, other features, that united the large hominoids to the exclusion of the gibbon and siamang. Among these was the "shutting down" of the two anterior palatine fenestrae. Then, within this group of large hominoids, there were two smaller groups.

One of these smaller groups consisted of the African apes, which were united especially because of the anatomical peculiarities related to these hominoids' mode of locomotion by knuckle walking. The other small group was composed of humans and the orang-utan, united because of their sharing a single incisive foramen, low-cusped cheek teeth, and thick molar enamel.

The fossils could then be put into place. Ramapithecids and fossil hominids belonged to the second group — the human–orang-utan group — because they, too, possessed these apparent evolutionary novelties of teeth and palate.

This scheme seemed to make sense. And it was internally consistent as well. The palatal distinctions seemed to have something going for them. The different configurations seemed to be distributed among the hominoids in evolutionarily meaningful ways. But this scheme also made a mess of what had come to be taken as the picture of hominoid phylogeny. After all, the African apes, if not just the chimpanzee alone, were supposed to be the most closely related among the hominoids to humans — no matter what the palate and teeth seemed to indicate.

I decided the next best thing to do was to get some feedback on this unexpectedly novel result, to see if what I took to be logical struck others as logical — that there were features, unique features, that yielded the hypothesis that humans and the orang-utan were evolutionary sisters. I wrote to my friend and colleague at the British Museum (Natural History) in London, Peter Andrews. Peter, a leading figure in the study of fossil hominoids, had published the first of the recent papers on the facial skeleton of *Sivapithecus* being orang-utan-like, and it was Peter who had provoked me into rethinking the evolution of hominoid primates. I sent Peter a diagram of my proposed theory of relationships among the hominoids with some shorthand notations on the features that would unite the various hominoid subgroups.

Peter wrote back that my human–orang-utan theory was logical and made sense of the problems that came from adhering to a human–African ape, and especially a human–chimpanzee, theory of relatedness. In fact, he wrote, he had also been reviewing the morphological data and could not delineate much at all that would suggest the unity of humans and one or both of the African apes. But he could not reject the popular theory of hominoid relationships, with humans being associated closely with the African apes, because this scheme seemed to be supported quite strongly by the molecular data, which he found convincing.

I guess I was a bit more skeptical than Peter about the conclusiveness of the molecular data. I had been reviewing that body of literature and had become bothered by the arguing that went on in those quarters, each proponent of a different biomolecular technique claiming that the approaches of others were inadequate to resolve phylogenetic questions. I was also becoming more entrenched in theoretical issues of reconstructing the evolutionary relationships of organisms. I was particularly intrigued by those instances when different types of data yielded markedly different, seemingly incompatible, phylogenetic schemes. This incompatibility was most apparent in the developing schism between molecular and morphological approaches to reconstructing the evolutionary relationships among the hominoids.

I published a short paper in the journal *Primates* on my preliminary thoughts on hominoid phylogeny — that "hominid" cheek tooth morphology and the possession of a single incisive foramen might reflect the relatedness of humans and orang-utans; that there did not seem to be very much from the study of morphology, or from the fossils, which held humans and the African apes together; and that we might bring the same questions of phylogenetic uncertainty to bear on the use of molecular data as the biochemists were charging against morphologists and paleontologists.

I had put my foot in it. It was now my obligation to see if my theory of relationships among the large hominoids and my questions about phylogenetic reconstruction and inference held water. I suspected that many colleagues would just reject my suggestions because they did not mesh with the going view of the world, but that would not be the same as providing evidence of contradiction or falsification. It was my theory — and thus, if I had learned anything at all from my colleagues in philosophy of science about being a scientist, I knew I had to try to knock it down.

7

Humans and
Orang-utans

INTELLECTUALLY STIMULATING though it might be to entertain the notion that humans and orang-utans are closely related, I knew that if I did not come back into line with the more popular notions of human–African ape relatedness, I would be in for a long siege. I would have to contend with morphologists and paleontologists who thought there were significant things to be discovered in the anatomy of humans and chimpanzees and gorillas — features that resulted from close evolutionary ties between our own species and these two apes. I would have to deal with those morphologists and paleontologists who also invoked molecular and biochemical research as the final arbiter of questions of relatedness. Lastly, I would have to deal with the molecular systematists and biochemists themselves, as well as those microlevel systematists who study chromosomes. Even though most of these researchers disagree about which of their approaches is the most accurate, they are fairly united in the belief that morphology cannot be considered a primary source in the unraveling of evolutionary relationships.

Since my theory of human–orang-utan relatedness was based on morphological considerations, I felt I should continue in a morphological vein. After all, it might well be that there really is a morphological basis for grouping humans with the African apes. Perhaps such evidence would turn out to be more compelling than that which seemed to support the grouping of humans and orang-utans. The matter would be resolved.

As I was gearing up for this project, a major review by Peter Andrews and Jack Cronin, a biochemist, appeared in *Nature*. It was entitled "The Relationships of *Sivapithecus* and *Ramapithecus* and the Evolution of the Orang-utan." It was Andrews's tour de force on the close affinities with the orang-utan of those fossils that had been earlier presumed to be related to the "proper" hominids, and Cronin's synthesis of the molecular data in support of the grouping of humans and the African apes. It was also Andrews's demonstration — on the basis of an extensive comparative study of craniodental features — that there was only one seemingly solid feature of the skull — the development of frontal sinuses — and a few other debatable ones, that supported the grouping together of humans and the African apes.

If you have ever had allergies or a sinus headache, you are aware of the spaces in the bone beneath the broader portion of your eyebrows and within your cheek bones. These air spaces are sinuses. Those in the bone behind your eyebrows are the frontal sinuses, while the sinuses in the cheek bones are the maxillary sinuses.

Frontal sinuses are not found in all mammals. Among primates, frontal sinuses are known to develop only in the large hominoids and do not develop with consistency nor in the same manner in all of the four large hominoids. The presence of frontal sinuses in these hom-

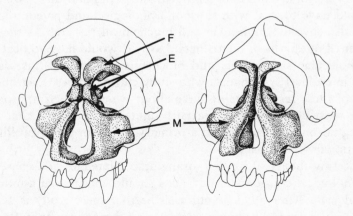

FIGURE 7A Facial sinuses in a chimpanzee *(left)* and an orang-utan *(right)*. Frontal sinus (F) development in the chimpanzee, as in gorillas and humans, is an outgrowth of the ethmoid (E). In the orang-utan, the maxillary sinus (M) may extend into the region of a true frontal sinus. In the chimpanzee the maxillary sinus invades the hard palate.

inoids cannot be thought of as a feature indicative of evolutionary closeness.

Orang-utans sometimes develop sinuses in the frontal bone, but these are not true frontal sinuses. "Frontal sinuses" in orang-utans are extensions of the maxillary sinuses, which have a tendency to become large and vacuous. These maxillary sinuses may be so expansive that they invade the interorbital region along the side of the eye. They may even continue burrowing their way up and around the top of the orbit, ending up in the area in which one finds a true frontal sinus in, for example, a human. In the chimpanzee, the maxillary sinus typically invades the bone of the palate, separating it into two layers.

Frontal sinuses in humans have a totally different origin. They arise from the sinus complex of a neighboring and very intricate bone located in the nasal cavity, a bone called the ethmoid. Two of its primary functions are to warm and filter incoming air.

The ethmoid bone is essentially a thin-walled, bony balloon covered with mucous membrane, laced with a network of capillaries, and filled with air. If you took a cross section of an ethmoid it would look vaguely like one of those statues of an Indian goddess with many arms. The ethmoid has three major "arms" on each side of its "body," each coiled loosely, and stacked one above the other. These bony coils or scrolls extend the length of the body of the ethmoid bone. In humans, the ethmoid sinus complex is composed of multiple air cells. The invasion of this ethmoid pneumatization into the frontal bone creates the frontal sinuses.

Chimpanzees and gorillas also develop frontal sinuses as a result of ethmoid sinus expansion. On this level of comparison, the African apes do indeed emerge as more similar to humans than any other primate, the orang-utan included. The comparison is so favorable that, earlier this century, the German comparative anatomist H. Weinert cited the development of frontal sinuses as a characteristic that almost single-handedly demonstrated the supreme closeness of humans and the African apes. Peter Andrews also took this feature to indicate the unity of these three hominoids.

Adolph Schultz, the Swiss primatological reincarnation of Thomas Huxley, argued strongly against Weinert's conclusions when they were first promulgated. He pointed out that frontal sinus development was much more variable in humans than in chimpanzees and gorillas.

Schultz also emphasized that within the species of chimpanzee, frontal sinuses are consistently developed only in the larger species, *Pan troglodytes*, the chimpanzee typically seen in circuses. The pygmy chimpanzee, *Pan paniscus*, rarely develops frontal sinuses.

The classic studies of A. J. E. Cave, a past president of Britain's prestigious Linnaean Society, demonstrate other differences among the hominoids in the configuration of the nasal cavity and its sinuses. For example, the frontal sinus of a chimpanzee or a gorilla, when it does develop, derives from a particular groove in the nasal cavity. In humans, frontal sinus development rarely originates from this structure. Chimpanzees and gorillas do not develop more than three ethmoidal air cells. In humans, this structure characteristically has multiple air cells or sinuses.

This discussion is not leading to the conclusion that there is nothing in the anatomy of the nasal cavity suggestive of close evolutionary ties between humans and one or both of the African apes. To the contrary, there are. But I don't think the link between humans and African apes — if one does exist — is to be found in frontal sinus development. There is too much variation and inconsistency in frontal sinus development among these three hominoids. If there is something in this general cranial region that can suggest the unity of humans and the African apes, it is that these three, and only these three, hominoids develop ethmoidal air cells.

Aside from the development of ethmoidal air cells, are there any other features that could support the grouping together of humans and the African apes?

In the earlier part of this century, the American vertebrate paleontologist and primatologist William K. Gregory attempted to demonstrate the closeness of humans with the African apes to the exclusion of the orang-utan. His methods left much to be desired. As I reviewed in earlier chapters, Gregory did not consider the possibility that the orang-utan might appear more dissimilar to humans than either of the African apes because the orang-utan had itself become different in the features being compared. Gregory also attempted to demonstrate the unity of humans with the African apes by the accumulation of comparisons that were specific to only one of the African apes. He probably assumed that the African apes constituted an evolutionary couplet, and felt that if one of these hominoids — either the chimpanzee or the gorilla — was similar to humans, the relatedness of both to humans was demonstrated.

Some of the features used are not really "kosher." For instance, Gregory cited such things as the humanlike "appearance of the pregnant female chimpanzee" and the development in old female chimpanzees and gorillas of pendent breasts. Gregory also compared the gorilla's nose with "the lowest existing types of human nose" and found them similar; in the same vein, Gregory found that the fundus of the eye of a chimpanzee is matched in detail by that of a "negro." Elsewhere Gregory felt that the closeness of humans and the African apes was indicated by humans and chimpanzees being similar in their "psychology."

Obviously we would reject these "demonstrations" of relatedness out of hand. They are a reflection of a notion of evolutionary transformation that is nothing more than the pre-nineteenth-century "great chain of being" veiled in Darwinian notions of selection. Even the more "morphological" of the suggested clues to phylogeny are now known to be inaccurate or just wrong.

For example, Adolph Schultz demonstrated that humans are unique among primates in developing a prominent nose. This nasal eminence exists because of the structural support of cartilage, not just as a strut from above, from which the nose can hang down, but especially in the "wings" of the nostrils, whose cartilaginous underpinnings allow for some mobility of the nostrils.

A broad human nose is caused by an expansion of the cartilage in the nasal wings. The "broad" nose of a chimpanzee and the even "broader" nose of a gorilla derive from the deposition of fat and the modeling of surrounding tissue. The acquisition of "a forward and downward growth of its tip," which Gregory had seen as the element needed to make a gorilla's nose human, will not transform the nose of a gorilla into the nose of any human. Since it is the enlarged nasal wing cartilage that makes the human nose what it is, and which distinguishes humans from all other animals, it makes sense to consider the broadest of human noses as being the most evolutionarily distinctive among humans.

Adolph Schultz amassed more data on the primates, especially on the hominoids, than anyone else before him and certainly anyone since. I believed his work would be a reliable source of comparative information that I could use to test the various theories of human–ape relatedness. Using Schultz's published data would, I thought, save me from years of redundant research and eliminate any possible claims of my having been biased in the collection of data.

From the 1920s until his last publications in the 1960s, Schultz studied virtually all cranial and skeletal aspects of primates. He studied body and limb proportions; counted ridges on palates and hairs per square inch of skin; reviewed sinus development, sexual differences, and changes that occurred during aging. His primary purpose was to gather information from which he could argue that the great apes were more similar to each other than any of them was to humans. Nevertheless, Schultz's data remain available for all of us to use and interpret.

In all of the hard- and soft-tissue anatomy Schultz scrupulously detailed, I could not find many features that humans shared with both of the African apes that were not also shared with orang-utans, gibbons, or monkeys.

If you run the tip of your tongue down the roof of your mouth, you will feel a series of ridges that course from side to side, between right and left sets of teeth. More prominent toward the front of your palate, they fade out as you get about a third to halfway back. Schultz discovered that in almost all humans these ridges are typically confined to that portion of the palate which lies in front of the second of the two premolar teeth. In some individuals, however, the palatine ridges may go as far back along the palate as the first molar.

If you were to peer into the mouth of virtually any other mammal, such as an opossum, a dog, or a horse, you would see a series of

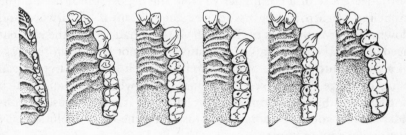

FIGURE 7B   The left half of the fleshy palate of *(from left to right)* an opossum, a gibbon, an orang-utan, a chimpanzee, a gorilla, and a human. The opossum has the primitive, typical mammalian pattern of palatine ridges: regular and along the entire palate. The hominoids, as a group, are different in that their palatine ridges are less regularly positioned and less pervasive in distribution. The large hominoids continue this irregularity even further. But the African apes and humans are most similar in confining the palatine ridges to the front of the palate.

somewhat parallel palatine ridges extending the full length of the palate — looking a bit like a stop-action photograph of a physicist's experimental ripple tank. This is the common mammalian condition.

If you were to look inside the mouth of a lemur or a monkey, you would see the typical mammalian pattern of rather uniform palatine ridges roughly undulating their way along the palate. Were you then to look into the mouth of a gibbon, you would see a somewhat different picture. The ridges would not be as uniformly distributed along the palate, and they would not extend the full length of the palate. The palatine ridges in a gibbon would appear to be more broken up and irregular.

The disruption of "orderly" palatine ridges and the truncation of their extent along the palate characterize all hominoids. But in contrast to the gibbon, the African apes, the orang-utan, and humans have even more "disorderly" palatine ridges. Orang-utans are similar to gibbons in that their palatal ridges usually extend as far back as the level of the second of the three molars. In the African apes, however, the palatine ridges do not usually go farther back in the mouth than the level of the first molar. Humans have the most extreme truncation of the palatal ridges, which often fade out before they even reach the first molar. If there is something potentially evolutionarily significant about the distribution of these different configurations of palatal ridges, I think it is in lending support to the human–African ape hypothesis of relatedness. The palatine ridges in these three hominoids are the most confined in their backward extent along the palate.

I was truly surprised to find only two seemingly clear-cut features — the development of ethmoidal air cells and the restriction of the palatine ridges along the roof of the mouth — that could be interpreted as indicating the evolutionary closeness of humans and the African apes. There is, however, a third, but shakier, possible feature.

Humans, chimpanzees, and gorillas are the only primates that occasionally develop ear lobes. However, ear lobe development is just not consistent enough a feature in these three large hominoids to support any phylogenetic notion of relatedness.

H. Weinert, Schultz's primatological adversary, had relied particularly heavily on two features to support the argument that humans were closely related to the African apes and not just to the great apes in general. One feature Weinert invoked to support this scheme was the development of frontal sinuses. Weinert's second major piece of

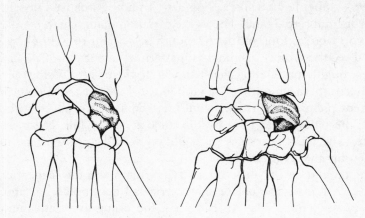

FIGURE 7 C   The right wrist of a typical Old World monkey *(left)* and a human *(right)*. The Old World monkey retains the primitive number of carpal bones in which the os centrale (stippled, *top*) and scaphoid (stippled, *bottom*) remain separate bones throughout the individual's life. In humans, as well as all other hominoids, these two carpal bones fuse to form a single wrist bone (stippled).

Another difference between hominoids and other primates is in the dissociation *(arrow)* of the ulna from its neighboring wrist bones. In the orang-utan and more so again in the gorilla, the lack of contact is the most marked.

evidence for the evolutionary unity of humans and the African apes came from developmental phenomena affecting the number of carpal (wrist) bones that a hominoid ends up with as an adult.

Most adult humans, chimpanzees, gorillas, and orang-utans have eight carpal bones making up the wrist joint. Adult prosimians and monkeys have nine carpal elements in their wrists. The fetuses of all primates, hominoid or monkey, have nine incipient carpal bones. What happens in humans and the African apes is that, not too long after birth, two of these carpal bones — the scaphoid and the centrale — usually fuse to become a single bone. Typically, these bones fuse together in the orang-utan, but such fusion is delayed until much later in life, sometimes as late as old age.

Weinert argued that early fusion in humans and the African apes of the scaphoid and the centrale was a reflection of the evolutionary closeness of these three large hominoids. The orang-utan, with its scaphoid and centrale fusing so much later in life, was more primitive than the others, and thus more distantly related to us than were the African apes.

Schultz, the great defender of the unity of the great apes, rejected Weinert's conclusion and evidence. Schultz pointed out that although humans and the African apes are different from monkeys in fusing the scaphoid and the centrale, and are different from the orang-utan in uniting these two carpals much earlier in life, early fusion was not restricted to humans and the African apes alone among hominoids. It is also characteristic of the gibbon. While fusion of the scaphoid and centrale at some time in life distinguishes hominoids from other primates, early fusion cannot be taken as reflecting the unity of humans and the African apes, precisely because it is also a feature in the development of the wrist of the gibbon. Once again, the orang-utan, because it typically fuses the scaphoid and centrale later than the other hominoids, stands out as being the most distinctive, and different, of the lot.

In this particular debate with Weinert, Schultz departed from his typical approach to evolutionary relationships among hominoids, which was to demonstrate that, overall, the large apes were more similar to each other than any of them was to a human. Instead, Schultz actually used the approach to resolving questions of relatedness I have been advocating. One cornerstone of this perspective is the determination of how the similarities among your taxa of immediate interest really are distributed throughout a broader sample. In this instance the similarity between humans, chimpanzees, and gorillas was not restricted to these hominoids alone. The gibbon, when included in the analysis, was found to be no different from humans and the African apes. The orang-utan became distinguished by its difference, by its dissimilarity.

I cannot stress strongly enough how important dissecting the nature of dissimilarity is to an analysis of evolutionary relationships. Evolutionary uniqueness can result in an organism's apparent dissimilarity just as much as remaining unchanged can, when compared to those organisms that have changed. An uncritical demonstration of similarity alone is not sufficient to warrant speculation of evolutionary relatedness.

Roughly coincident with the publication in the journal *Primates* of my paper on palatine fenestrae, incisive foramina, and the possible relatedness of humans and orang-utans, a few papers on other aspects of the nasal cavity appeared which seemed to contradict my findings.

Steve Ward, Bill Kimbel, and David Pilbeam, in various combinations as co-authors and collaborators, had studied the floor of the
nasal cavity of various living and fossil hominoids. They concluded
that there were two different configurations characteristic of two different groups of hominoids. One configuration they called the "African pattern." It indicates the relatedness they propose of the
chimpanzee and *Australopithecus*. The other configuration of the floor
of the nasal cavity they identified as the "Asian pattern," which is
supposed to reflect the unity of the orang-utan and its apparent fossil
ally, *Sivapithecus*.

In a review for the widely read American journal *Science* of a
volume that contained one of these articles on the "African" versus
the "Asian" pattern, Eric Delson commented on the apparent disparity
between their conclusions and mine — conclusions seemingly derived
from study of the same anatomical region. The fact of the matter,
however, is that our descriptions are compatible. Ward, Kimbel, and
Pilbeam concentrated on the region of contact between the premaxilla
and maxilla in the floor of the nasal cavity, *within* the nasal cavity.
I had focused on what was happening on the oral-cavity side of the
palate. We were working within centimeters of each other but on
opposite sides of the palate, studying different details of different
expressions of the same basic region. Yet even if we had been studying
the same thing we would probably have come to different conclusions
about who was related to whom. Where we seem to disagree most is
in the interpretation of the similarities we discover.

The general region in question is the same one discussed in chapter
6, the region that may be perforated by palatine fenestrae (those
"windows" that communicate between the nasal and oral cavities),
or by close-together palatal canals, or by a single incisive foramen.
The canals appear to be created by the "shutting down" of the large
palatine fenestrae. The presence of two long midline canals is distinctive, for instance, of the gorilla. In a more derived version of
the "shutting down" of the palatine fenestrae, the two canals merge
into one bony "hole" as they course from the nasal cavity toward
the oral cavity. This single palatal canal, the incisive foramen, is
broad and deep in humans and other hominids. Orang-utans also
have a single incisive foramen, but it is much more compressed and
slitlike.

The palatine fenestrae, since they are "holes" in the palate, disrupt complete contact between the premaxilla, the bone at the front of the upper jaw, and the long, bony palate behind it, which is part of the maxilla. In most mammals, the floor of the premaxilla and the bony palate are relatively thin. The palatal region of a large hominoid is a much thicker chunk of bone. The premaxilla and the maxilla behind are thicker and deeper and the contact between them more complete and continuous. It is important to remember that all hominids, and I mean all species of *Australopithecus* and all species of *Homo*, have a single incisive foramen perforating the palate. But from the least severe configurations of "shutting down" the palatine fenestrae, as seen in the gorilla, to the most severe, as seen in the orang-utan, there will always be two openings in the nasal cavity, one on either side of the vomer, the perpendicular plate of bone that bisects the nasal cavity.

In the African apes, the contact between the premaxilla and the maxillary bony palate behind is somewhat slanted, but the backward inclination of the premaxilla is benign enough that the posterior and upper margin of this bone does not greatly overlap the anterior edge of the maxilla. Where it does overlap the maxilla, the posterior and upper margin of the premaxilla may be slightly thickened and thus raised. This produces, depending on the African ape and the individual under study, a "stepping down" to the floor of the nasal cavity. In the African apes, the entrances of the canals on either side of the vomer can be quite large, like the mouth of a funnel. These canals are relatively larger and the development of "stepping down" more pronounced in the gorilla than it is in the chimpanzee.

The nasal cavity of all species of *Australopithecus*, but especially *Australopithecus afarensis*, approximates the "African pattern," particularly as it is expressed in the chimpanzee. From this level of similarity between one of the African apes and particularly one species, *afarensis*, of one genus of hominid, *Australopithecus*, Pilbeam and his colleagues have concluded that, in general, the "African pattern" demonstrates the overall relatedness of hominids and the African apes. I disagree from the start with the evolutionary interpretation of the "African pattern" of the nasal cavity.

Regardless of similarities, the species of *Australopithecus* differ distinctly from the chimpanzee, as well as from the gorilla, in that they typically develop a single incisive foramen palatally. This distinguishes

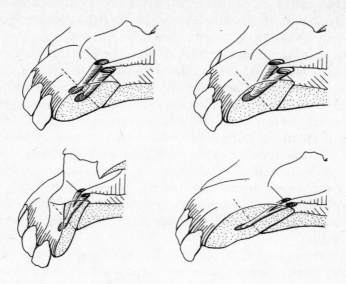

FIGURE 7D The anterior portion of the "snout" of hominoids, illustrating schematically and in section the nature of the contact of the premaxilla (with teeth) and maxilla behind it, as well as the configurations of the canals that course between the nasal and oral cavities along the contact zone of the premaxilla and maxilla. The upper left figure illustrates the "African pattern," characteristic of the chimpanzee and gorilla, in which the contact between the premaxilla and maxilla is relatively vertical; two canals, retaining their separateness, are also present. The upper right illustration represents the condition hypothesized to have characterized the last common ancestor of humans and orang-utans; the general configuration of the premaxilla and maxilla typical of the "African pattern" is retained, but what is unique here is the confluence of the two bony canals as they course toward the oral side of the palate to form a single incisive foramen. From the configuration of premaxilla, maxilla, and canals that would have characterized the last common ancestor of humans and orang-utans (and which would have been retained in *Australopithecus*), humans *(bottom left)* and orang-utans *(bottom right)* became differently derived. In humans, the premaxilla is short and thick and forms a very steep contact with the maxilla behind it; perhaps as a consequence of this palatal/facial "reorganization," the canals are narrow but the single incisive foramen remains large. In the orang-utan, the premaxilla is elongate and posteriorly overrides the anterior extent of the maxilla, forming an oblique contact; perhaps as a consequence of this configuration, the canals are narrow and their confluence emerges on the oral side of the palate as a long, slitlike single incisive foramen. The details of the premaxilla and its contact with the maxilla have been referred to as the "Asian pattern," which, along with the details of the canals and single incisive foramen, also characterizes *Sivapithecus*.

Australopithecus and *Homo* from both of the African apes. The most that *Australopithecus* shares with the African apes is the consequence in the nasal cavity of "shutting down" the palatine fenestrae — the reconfiguration of these perforations of the palate into canals on either side of the vomer. The African apes do not share with *Australopithecus* the additional element of the two canals typically merging into a single incisive foramen palatally. *Australopithecus* shares this feature with *Homo*, the orang-utan, and *Sivapithecus*.

During my study of the nasal and palatal regions in primate skulls, I tried to pass bristles — actually, thin plastic bristles from the brush end of an eraser pencil — through the various openings and canals in order to trace their paths and interminglings. It was easy to "thread the needle" when sticking a bristle into the openings in the nasal cavity of a chimpanzee skull; the funnellike aperture helped guide the probe. Gorilla skulls, because of the larger "mouths" of their canals, were sometimes easy to probe, but the thickened posterior and upper margin of the premaxilla often interfered with an easy and direct thrust of the bristle. With modern human skulls one had to be right on target to get that bristle into the opening of a canal, because the canal is much narrower and much more appressed to the base. The reduced size of the canals in humans may be correlated with the fact that the contact between premaxilla and maxilla in humans is more vertical than in any of the apes, and the posterior and upper margin of the premaxilla in humans is expanded as a sheet of bone that overrides the anterior upper edge of the maxilla.

The "Asian pattern," seen in the orang-utan and the fossil *Sivapithecus*, is characterized by a longer and markedly slanted contact between the premaxilla and the maxilla. The anterior and lower edge of the bony palate projects further forward, underriding the premaxilla. The posterior and upper edge of the premaxilla projects more extensively backward, overriding the maxilla. The development in the orang-utan and *Sivapithecus* of a long, slitlike single incisive foramen is probably related to this long, slanted contact. We know that, of the two, the orang-utan at least also has very narrow nasal-cavity canals, which may have become reduced in size as a result of the development of an oblique zone of contact between the premaxilla and the maxilla. Given the difference in detail of this contact zone in humans and orang-utans, as well as the persistence of larger nasal-cavity canals in fossil hominids, we must conclude that modern humans and orang-utans have independently reduced the size of their nasal-cavity canals.

I do not think that both patterns of the nasal cavity, the "African" and the "Asian," can be phylogenetically revealing with regard to the unity of humans and African apes as well as the unity of the orang-utan and *Sivapithecus*. Each pattern can, of course, be derived. But I think that each pattern is derived at a different level within Hominoidea.

The "African pattern," as seen particularly in the chimpanzee and *Australopithecus*, would have characterized the last common ancestor of all the large hominoids. The presence of the "African pattern" in the African apes and *Australopithecus* is due to the retention of it from these hominoids' past ancestor. A more recent ancestor would have been distinguished by its development of a single incisive foramen palatally. This ancestor would have been the last common ancestor of the hominids (which includes all species of *Australopithecus*), the orang-utan, and *Sivapithecus*. Modern humans would have become yet different again by reducing nasal-cavity canal size and by thickening and straightening the contact between the premaxilla and the maxilla. The "Asian pattern" of the nasal cavity and the long, slitlike single incisive foramen would have taken shape with the emergence of the last common ancestor of the orang-utan and *Sivapithecus*.

For whatever reasons — repetition, the weight of authority, or both — the "African pattern" versus "Asian pattern" argument for resolving the relationships among the large hominoids has begun to find a place in paleoanthropology. I am not among the convinced. At the level of distinguishing evolutionary groups, the "African pattern" may have characterized the last common ancestor of the large hominoids, but only the "Asian pattern" delineates a smaller group within the assemblage of large hominoids. What remains after uniting *Sivapithecus* and the orang-utan because their nasal cavities are configured along the lines of the "Asian pattern" is not a human–African ape alliance.

Since it is becoming increasingly popular these days to accept the notion that the chimpanzee is the most closely related to humans, I thought I would try to see what Adolph Schultz had in the way of data unique to these two hominoids. The only possible connection I could dredge up from his sources dealt with hair density. In yet another of his scrupulously compulsive studies, Schultz counted the number of hairs per square centimeter on the scalp, back, and chest of representatives of each of the hominoids as well as some specimens of other primates. He discovered that humans and chimpanzees are distinctive in having the fewest number of hairs per square centimeter on their backs.

Essentially, then, these bits and pieces of information — Schultz's data on palatal ridges and hair density and Cave's analyses of the ethmoidal air cells of the nasal region — are what I came up with in support of either a human–African ape or a human–chimpanzee theory of relatedness. In his *Nature* article with Jack Cronin, Peter Andrews suggested that, in addition to the development of frontal sinuses, humans could be united with the African apes on the basis of the thickening and enlarging of the area over the eye sockets (the supraorbital region) as well as the swelling in the region above the bridge of the nose. These are interesting features to consider, but, as Peter admitted, the development of prominent brow ridges and their tendency to come together above the nose is more characteristic of fossil hominids than it is of modern *Homo*. If the development of these supraorbital structures really does unite hominids and the African apes, we are forced to conclude that such features must have been lost with the emergence of our own species.

Perhaps Peter is correct, and aspects of brow ridge development do unite humans and the African apes. If we extend the comparison to orang-utans, we find that their supraorbital region is very smooth and gracile. Gibbons and siamangs also develop brow ridges and swollen supranasal (glabellar) regions, but these developments are benign compared to what one finds in chimpanzees and gorillas and many of the fossil hominids. If we look even farther afield from the hominoids, to the Old World monkeys, we discover that brow ridge development is quite common. In the highly sexually dimorphic African apes, these supraorbital and supranasal embellishments can be quite marked in the male. It might be that enlargement of the brow ridges and swelling of the glabellar region are not specific enough to the African apes and hominids for them to be considered features indicative of the evolutionary closeness of these three hominoids.

Perhaps such cranial distentions would have been distinctive of the ancestral catarrhine, the common ancestor of Old World monkeys and hominoids. The presence of these features in the African apes and many fossil hominids would be the result of the retention of primitive craniofacial features. The relative underdevelopment of these regions in gibbons and siamangs and especially in orang-utans and modern humans would represent the unique cases.

Although Schultz's data are surprisingly skimpy on features that could potentially be interpreted as uniting humans and one or both of the African apes, his data are unexpectedly replete with features shared uniquely by humans and orang-utans.

For example, and while we're not too far removed from the topic of hair, Schultz discovered that "man and orang-utan are the only higher primates which can produce excessively long hair." "In an adult male orang-utan," Schultz proclaimed in formal fashion, "the author found hair on various parts of the body exceeding 55 cm. in length, a dimension equalled only by the scalp hair of some human beings."

In another study, Schultz analyzed the shapes and relative proportions of the subunits of the scapula (shoulder blade). It was in that study that Schultz pointed out that the large hominoids are distinguished from other primates in having relatively shorter and deeper scapulae. Most mammals have an elongate, narrow scapula, somewhat triangular or fan-shaped, depending on how rounded the posterior vertebral border is. Since most mammals are quadrupedal, the scapula lies along the side of the rib cage, with its long axis oriented almost vertically. The narrower end of the scapula swells slightly to form a modest cup, which sits like a shallow helmet on top of the humerus, the bone with which it articulates. The broad end of the scapula points toward the vertebral column.

Although the shape of the scapula of most primates is not dissimilar from the shape of a mammalian scapula, there are noticeable differences. The lengthwise ridge, which on your own scapula you can feel running horizontally toward your shoulder joint, is more pronounced in primates. The scapula of a primate bears two prominent projections in association with the cuplike end of this bladelike bone. One projection is an extension of the scapular ridge which overrides the top of the humerus and with which the end of the collarbone comes into contact. You can feel how the spine continues forward and comes to a corner over the humerus, where it is abutted by the clavicle. The other scapular projection is a small knob just to the inside of the head of the humerus. You have located it when it hurts from being poked too much.

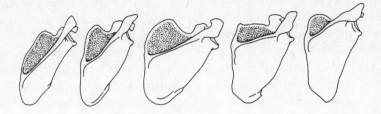

FIGURE 7E The right shoulder blade, or scapula, of *(from left to right)* a gibbon, a chimpanzee, a gorilla, an orang-utan, and a human; the humerus articulates with the socket on the right of each scapula. In humans and orang-utans, the posterior side of the scapula, which is closest to the vertebral column, is the straightest and the spine of the scapula, which delineates an upper (stippled) muscle attachment region from a lower one, is most horizontal. Also characteristic of humans and orang-utans, the upper region of the scapula is quite reduced in size relative to the lower portion.

While the large hominoids, as a group, may be distinguished among primates by their having relatively shorter and deeper scapulae, humans and orang-utans are more distinctive in having by far the relatively shortest and deepest scapulae. Humans and orang-utans are further distinguished in having a scapular spine that is oriented so horizontally that it nearly forms a right angle with the relatively straight vertebral border of the scapula.

Between humans and orang-utans and other primates there are also differences in the relative proportions of certain definable regions of the scapula. In humans and orang-utans, the area of muscle attachment above the scapular spine is quite reduced compared to the segment of the scapula that lies below the spine. More typically among mammals, the scapular spine runs closer to the midline of the blade — not quite bisecting the blade, but not creating a scapula that appears grossly asymmetrical from top to bottom. You can feel all of this while massaging a dog's or cat's shoulders and can compare this configuration with the greater asymmetry you will feel on your own — or, probably better, on someone else's — scapula. The area above your scapular spine is quite small compared to the area below the spine. And it is always the small suprascapular muscles that give one the most trouble.

Recently, Elisabeth Vrba, a paleontologist formerly at the Transvaal Museum in Pretoria but now at Yale University's Peabody Museum, who usually studies antelopes and other bovids, published a report on a scapular fragment of an *Australopithecus* from the site of Sterkfontein, South Africa. Vrba found that the shallow scapular socket, which articulates with the humerus, was deflected at an angle similar to that characteristic of the gorilla and the orang-utan. This seems curious when you consider that the gorilla is the least habitually arboreal and the orang-utan the most habitually arboreal of the large hominoids. An upturned scapular head would seem to be correlated with suspensory, arm-hanging types of arboreal locomotion. But perhaps it is not all that strange after all. Among the large hominoids, the lower end of the ulna is the farthest removed from articulation with the carpals in the gorilla and in the orang-utan. This would give greater rotation at the wrist, which should be beneficial for an active arborealist, either present-day or ancestral.

Vrba also discovered that the disposition of one of the two scapular processes — the one that presents itself as a small bump in front of the head of your humerus — conformed to a pattern otherwise seen only in *Homo* and the orang-utan. This is definitely of interest in terms of the potential evolutionary relationships of humans and orang-utans.

If we linger a bit longer in the arm and shoulder region, there is additional information. Think back to the discussion on delayed ossification of epiphyses of the various bones of the arm. All hominoids, large as well as small, are distinguished from other catarrhines by a delay in the onset of ossification of the lower end of the humerus as well as the epiphysis with which it articulates, the upper end of the radius, which is the lower arm bone that rotates when you turn your palm up. All hominoids also retard the onset of ossification of the epiphyses of the first set of finger bones and of many of the carpal bones. Within Hominoidea, the large hominoids are set apart from the gibbons by their delaying ossification in the lower epiphysis of the ulna as well as in part of the upper epiphysis of the humerus.

But humans and orang-utans are even more distinct in that they retard the onset of ossification in yet other epiphyses: the second and larger of the two upper humeral centers of ossification and the lower end of the radius.

I have tried to amass information on other parts of the skeleton, but have found that much more difficult than I had anticipated. Part of the problem is that a lot of comparative work still needs to be done, and published. But another part of the problem is that it has only been in recent years that the orang-utan has begun to receive the attention that it deserves.

As part of an intensive course I was trying to give myself on hind-limb anatomy, I attended a full-day symposium on the evolution of the hominoid foot during the annual meeting a few years ago of the American Association of Physical Anthropologists. There was considerable debate on the functional interpretations of Lucy's foot. Did she walk upright and bipedally? Did she walk with a stride? Or were her toe bones too long and curved for her to have been a habitual biped? Was she, instead, most adept at shinnying up and down trees? The room was hot and stuffy. Then Henry McHenry got up to give his paper.

Henry is a miracle among paleoanthropologists. I have never heard him utter a nasty comment about anyone, and I have never heard anything unseemly said about him. With all the controversy and heated debate that spews almost daily from the centers of paleoanthropological research, Henry has somehow managed to thread his way around it all. On top of that, McHenry Vineyard produces a great pinot noir.

Henry's standard opening with a funny line brought me out of a dull, "I've-been-in-this-room-too-long" haze. Then toward the middle of his short presentation, he commented in passing on how humans and orang-utans were different from other hominoids in a particular aspect of one of the large ankle bones, the talus.

Henry's newly found piece of anatomical information was not much — just that a tubercle, basically a bump, on the talus is reduced in humans and orang-utans. In the chimpanzee and the gorilla this talar tubercle is long, which may be its more common disposition among catarrhine primates. This bit of data, in and of itself, is not a whole lot to go on. But I think it might serve as an indicator of a broader array of detail that would be uncovered if a more intensive comparative study of the postcranial skeleton were undertaken.

If we go back to the other end of the body, to the skull and teeth, there are things to be considered. It was from the study of the teeth — aspects of cheek tooth cusp height and molar tooth enamel thickness — that the present upheavals in the deciphering of human ancestry arose. It was also because humans and orang-utans were found to be similar in specific details of their teeth, with premolars and molars having low cusps and molars being capped with a thick layer of enamel, that I became interested in the first place in matters having to do with hominoid evolution. I just could not assume from the beginning that there was nothing significant about such seemingly unique similarities.

With regard to other aspects of dental morphology, however, comparison beyond cusp height, enamel thickness, and general tooth shape is confounded by the fact that the surfaces of an orang-utan's premolars and especially its molars are riddled with creases, grooves, and crevices. This condition is referred to as "wrinkled" enamel. This elaborate, yet haphazard, network of surface detail virtually obliterates the identity of telltale crests, ridges, and various small cusplike structures that are the focus of comparative dental studies.

During the course of gathering data on the hominoids, I was often struck by how much of the interpretation of the evolutionary significance of such information was predicated on humans' being closely related to the African apes. Sometimes I think I would not have become so involved in this area of paleoanthropology at all if it were not for the overriding authority this particular view of hominoid evolution wielded — over almost everything. The more something has become an evolutionary "truth," the more inclined I am to try to find out what makes it so.

There are certain colleagues whose articles I always try to read and whose presentations at meetings I always try to hear. Steve Ward, of Kent State, who has worked with David Pilbeam on various *Sivapithecus* problems, is one of these colleagues. At a recent annual meeting of the American Association of Physical Anthropologists, Steve gave a paper on features of the hominoid upper canine — basically, on how the upper canine teeth of hominoids, fossil and extant, were implanted differently in the jaw in different taxa. For example, in such fossil forms as the Miocene *Proconsul* as well as in the chimpanzee and the gorilla, the stout root of the upper canine is pretty much in line with the crown of the tooth. Perhaps this does not strike you as anything to comment on, but this configuration is not found in all hominoids.

In one of the many studies on *Australopithecus* teeth published in the 1970s, it was pointed out that the crown of the upper canine of these hominids and the root of the tooth are not in the same plane. The crown of the upper canine of *Australopithecus* is rotated relative to the root of the tooth. The configuration is perhaps best envisioned as resulting from holding the tip of the canine's thick root giving the tooth a slight twist. Upper canine rotation is a feature that came to be associated with being hominid.

As a result of Pilbeam's remarkable history of discovery of new *Sivapithecus* material from Pakistan, Ward could study upper canine rotation in this hominoid. And *Sivapithecus* had it — the slight twist to the upper canine seen in hominids, not the aligned canine root and crown one sees in *Proconsul* or the extant African apes.

In the "good old days" before *Sivapithecus* emerged as being in distinctive morphological detail most similar to the orang-utan, this discovery of upper canine rotation would have been greeted with open arms as yet another feature that reflects the relatedness of the Miocene form with later hominids.

But we know now that *Sivapithecus* is related to the orang-utan. And because the orang-utan is not supposed to be closely related to hominids, upper canine rotation cannot be phylogenetically significant within Hominoidea and certainly cannot be meaningful when it comes to hominid phylogeny.

This conclusion would seem to be further warranted because Steve discovered that orang-utans typically have some rotation of the upper canine. The twist may not be as profound as in *Sivapithecus* and *Australopithecus*, but it is usually there. (I say "usually" because recently Bill Kimbel pointed out that a slide I showed of the upper jaw of an orang-utan did not illustrate canine rotation. I guess I had found the exception to the rule.) Given the supposed phylogenetic separation of orang-utans from hominids, upper canine rotation could not, ever again, be perceived as having any phylogenetic relevance in the realm of human evolution.

That evening, as so often happens during meetings where the hotel is the only place to be, I found myself, with Steve Ward and a few others, crammed around a tiny but clunky table in one of those neo-antique hotel bars where you can hardly hear yourself think and end up getting more beer spilled on you than you can get down your throat.

I did manage at some point to ask Steve if he had considered the possibility that the reason he found upper canine rotation in *Siva-*

pithecus as well as in the orang-utan and hominids was because these hominoids in particular were closely related. He hadn't, he admitted, because he had accepted the close relationship of the African apes to hominids. That was that.

I have learned, although much too often the hard way, not to preach too much in the hope of achieving instantaneous conversions. The subject was dropped and our discussion turned to how we were going to overcome our shared fears of height and exposure so that we could get into one of those awful glass elevators and make it to the top floor of the hotel. But before entering that tiny torture chamber, I did make a mental note about Steve's finding the otherwise apparently hominid feature of upper canine rotation in *Sivapithecus* and the orang-utan.

In the course of reviewing the literature for data that could be used in reconstructing the relationships of the hominoids, I reread an old article by Eric Delson and Peter Andrews on some comparative aspects of the skull and teeth of catarrhine primates. I remembered this article because Delson and Andrews had expressed doubt as to exactly what features held humans and the African apes together. A few years later, in a co-authored volume on primate evolution, Delson would admit that the linking of humans with the African apes has been based largely on similarities that are primitive retentions — features not particularly representative of, nor specific to, humans and African apes alone. Phylogenetic interpretations aside, an important feature of Delson and Andrews's work is that they provide data — tables of it, in the case of the earlier joint article.

In going over their lists on various dental and cranial features, I was surprised to find that there were no features that could be construed as indicative of the unity of humans and the African apes. But there were two features shared by humans and orang-utans which Delson and Andrews had identified as being derived — that is, evolutionarily novel. One of these was the possession of a third lower molar, or "wisdom tooth," smaller than the molar in front of it. This size relationship is characteristic of all species of the genus *Homo*. However, some *Australopithecus*, especially the more robust and massive species, have a third lower molar that is the longest of the lower molars.

Given the size of the third lower molar in robust *Australopithecus* and the fact that the African apes and the gibbons and siamangs also have relatively large third lower molars, is the comparison between humans and the orang-utan evolutionarily significant? Some colleagues object to my citing small third lower molars as a feature that may unite humans with the orang-utan. Their objection is based primarily on the negative evidence presented by the robust species of *Australopithecus*. And given that the posterior molars of *Gigantopithecus* and *Sivapithecus* are less unequal in size, it might be that the smaller third molars of humans and orang-utans do not reflect any particular closeness between these two extant hominoids.

The other dental feature gleaned from Delson and Andrews's tables was the absence of a particular elaboration of the enamel of the upper molars of humans and orang-utans. When this enamel buildup is present in the molars of other mammals, it is frequently on the lingual (tongue) side of the tooth.

If you run your tongue along the inside of your upper molars, you will feel nothing of particular morphological note. You should feel vertical grooves along the side of each molar, but these grooves are the natural delineations between the cusps of the tooth. If you are good at sensing things with your tongue, you might be able to identify the fourth cusp, the one that "squares up" the corner of at least the first of the upper molar teeth.

Things would be different, however, if you were a *Proconsul*, one of those Miocene fossil hominoids from East Africa. You would actually feel something of consequence as you ran the tip of your tongue along the lingual surfaces of your upper molars: you would feel a thick ridge of enamel, called a cingulum, extending around the inner surface of each molar. And the surface of this thick ledge of enamel would be wrinkled, so wrinkled that its surface would look as if it were adorned with a string of minuscule irregular beads.

Most of the fossils that have ever been thought of as potentially hominoid have had thick upper molar cingula. Among the extinct hominoids, only *Sivapithecus* and *Gigantopithecus* did not develop this thickened ledge of enamel on their upper molars. Among the extant hominoids, gibbons usually develop fairly marked cingula on all three molars. Gorillas and chimpanzees develop upper molar cingula, but it is typically not expanded and ledgelike and it usually does not completely surround the perimeter of each molar. Humans and orang-utans are even further removed from the *Proconsul* or gibbon state of cingulum development. The upper molars of humans and orang-utans are generally devoid of any buildup of cingulum anywhere.

Before I began organizing my thoughts for this discussion of upper molar cingula, I had not really paid much attention to the ways in which cingulum development in the African apes varies or how the layer of enamel is reduced in thickness. Now that I have begun to put this data into a broader context, it strikes me that this feature may have been characteristic of the last common ancestor of the four extant large hominoids. And that novelty would have been the developmental breakup, inhibition, or retardation of the primitive configuration of having thick, ledgelike upper molar cingula. The last common ancestor of humans and the orang-utan would have gone even farther in retarding or inhibiting altogether the development of upper molar cingula. And it was from this ancestor — an ancestor not shared by the African apes — that modern humans, fossil hominids, the orang-utan, *Sivapithecus*, and *Gigantopithecus* inherited their smooth-sided upper molars.

In contrast to the permanent teeth, one can squeeze more minute morphological detail from a study of the deciduous, or "milk," teeth. The wrinkling of these earlier teeth is not as extensive as it is in adult

FIGURE 7 F   Upper left second molar of *(from left to right)* a gibbon, a gorilla, and an orang-utan, viewed from the lingual (tongue) side of the tooth, looking across the occlusal (chewing) surface. In the gibbon, a thick band of enamel, called cingulum, typically embraces the lingual side of the upper molar. In the gorilla and the chimpanzee as well, upper molar lingual cingulum is often less developed. In the orang-utan, as in humans, upper molar lingual cingulum is absent.

Looking across the occlusal surface, cusps and crests between cusps are more noticeable and clearly delineated in the gibbon and gorilla (as well as the chimpanzee) than they are in the orang-utan (and humans). Lower cusp height and rather obliterated crests in orang-utans and humans are actually related to the development of a thicker layer of enamel than in gibbons, gorillas, and chimpanzees. In the gibbon and gorilla illustrated here, the enamel has worn through on the tips of most molar cusps, whereas they are still intact in the orang-utan. Note that the enamel of the orang-utan's molars is extremely wrinkled.

teeth, there are fewer teeth involved, and, as their name indicates, the deciduous teeth will eventually be shed. These minor disadvantages aside, J. Douglas Swarts, one of my graduate students, pursued a broad and detailed comparative study of the deciduous teeth in anthropoid primates — New and Old World monkeys as well as the hominoids. He came up with some interesting information. In the shape of especially the first milk molars in the upper as well as lower jaw, and in specific details of their crests, cusps, and intervening basins, the milk teeth of humans and orang-utans are distinctly similar and, as such, distinguished from the teeth of other anthropoids.

After I sent my manuscript on palatine fenestrae and incisive foramina to the journal *Primates*, I returned to the study I had been working on before I got sidetracked by hominoids. I set up all those skulls again — cranial bases and teeth facing up — and resumed my attack on the problem of tooth and bone interaction. Once again, my mind wandered.

I found myself staring at the bases of those skulls, staring at and comparing the various holes and fissures that perforate the dense, rugged bones that are all jammed together into such a small but important area of the cranium. Many of these cranial "holes," or foramina, are small, permitting passage of only the more slender nerves or blood vessels, but there is always a large foramen, the foramen magnum, through which the spinal cord snakes its way from the brain toward the tip of the sacrum.

Perhaps because the skulls were all oriented the same way, with their palates, teeth, and craggy bases facing upward, I was reminded of an illustration in a publication on *Australopithecus afarensis* by Tim White, Don Johanson, and Bill Kimbel. In this article, two facing pages (and these were large pages) were filled completely with drawings of the left basicranial regions of various *Australopithecus*, a modern human, and a chimpanzee. The primary purpose of this illustration was to demonstrate the overall similarity of *Australopithecus afarensis'* basicranial region and the chimpanzee's, thus allowing one to argue for the primitiveness of the fossil hominid *A. afarensis*. If a hominid was indeed so primitive, then there was the possibility that this hominid was ancestral to all other hominids.

These sorts of unspecific comparisons make me a bit uncomfortable, so I wondered if any hominoid other than a chimpanzee would be as generally similar to *Australopithecus afarensis*. Sure enough, the gorilla and the orang-utan had equally similar basicranial regions. Even the more specific area of the base of the skull, the mastoid region, to which Johanson and colleagues drew special attention, was the same in all three apes, and thus compared well to the mastoid region of *Australopithecus afarensis*. In the fossil hominid and the three apes, the mastoid region, which you can identify on yourself as a large and downwardly distended bony projection behind and below your ear (or ear lobe), is quite flat and relatively unsculpted. Muscles that rotate your head stretch from the top of the sternum (breastbone) to the mastoid region. A curious aspect of the article by Johanson and his colleagues is that although the illustration isolates the chimpanzee for comparison to *Australopithecus afarensis*, you can pick carefully through their text and find that they do realize that this region is similarly flat in all of the apes.

Another area of the basicranium that was singled out as being particularly similar in the chimpanzee and *Australopithecus afarensis*, from which other *Australopithecus* as well as *Homo* differed, was the dense bone of the ear region. This bone is so thick and craggy that it is called the petrosal bone, which derives from the Greek word for rock, *petrus*. The petrosal bone houses the semicircular canals, which function to maintain an animal's balance, as well as the three tiny middle ear bones, which are located just on the inner side of the eardrum. The petrosal bone continues beyond the eardrum as a thick, bony tube that courses to the side of the skull and whose opening becomes confluent with the fleshy canal of the outer ear.

The foramina that perforate the human petrosal bone are huge. There is a massive foramen into which the internal carotid artery, after parting company with the large common carotid artery on the side of the neck, will proceed. Once inside the petrosal bone, the internal carotid artery will send a branch, rich in oxygenated blood, to the brain. The reason you can sometimes hear your heartbeat at night and can feel your pulse on the side of your neck is that the common carotid arteries come right off the aortic arch, which, in turn, comes right off the heart. There is virtually a straight connection between your heart and your ear region.

Also associated with and formed in part by the petrosal bone is a large foramen for the jugular vein, which is the vessel that takes oxygen-stripped blood away from the brain. And just in front of and bordered by the tip of the petrosal bone — that bit of the petrosal

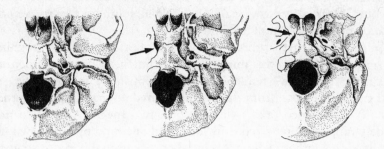

FIGURE 7G Basal view of the skull of a chimpanzee *(left)*, an orang-utan *(center)*, and a human *(right)*. (All are drawn to the same scale.) In general, the basicrania of the chimpanzee and the orang-utan are similar and remain primitive compared to the somewhat remodeled human skull. However, in terms of specific, unique features — such as the development of a foramen lacerum *(arrows)* — the human and orang-utan basicrania are together set apart.

that extends the closest to the midline of the basicranium — there is a large, jagged foramen, the foramen lacerum, whose edges actually do look somewhat lacerated. Unlike the other foramina of the skull, nothing passes through the foramen lacerum. This foramen is plugged up with fibrous tissue and cartilage. Nevertheless, humans are distinguished by their having a foramen lacerum.

In White, Johanson, and Kimbel's illustration of various hominid basicrania, only the modern human skull was portrayed — correctly — as having a foramen lacerum. This foramen was not present in the chimpanzee and the specimens of *Australopithecus* illustrated.

I looked at the skulls in front of me.

It was true — virtually every primate skull lacked this foramen. Except the orang-utan. It was tiny, to be sure, but a foramen lacerum was present, toward the midline of the basicranium, just near the tip of the petrosal bone. To be even more certain about this "discovery," I plowed through all my reference books and pored over all the photographs and drawings of basicranial regions I could lay my hands on. My "discovery" seemed to be borne out. Except for humans, the only anthropoid that had a foramen lacerum was the orang-utan.

A few weeks later I went to New York City to do some research in the mammal collections of the American Museum of Natural History. Before I got caught up in the orang-utan controversy, I had

begun a quiet little project on the taxonomy of the slow loris, which
is found throughout the Malay peninsula, southern China, and the
islands of Borneo, Sumatra, and Java.

In the early part of this century, primate taxonomists recognized
dozens of species of this mink-sized prosimian. But over the decades
the more "modern" tendency of taxonomists to lump species together — to remove from the overpopulated animal kingdom unnec-
essarily subdivided taxa — left us with two species, one for a now
widely ranging mink-sized slow loris, and one for a pygmy slow loris,
confined to southern China. I think there are probably six or so good,
valid, recognizable species of the larger slow loris. And one place to
go at least to begin to investigate the problem is the American Museum
of Natural History, which has one of the best collections anywhere
of this particular prosimian.

It is always refreshing to return to New York City to the museum's
dimly lit, echoey halls lined with collection cabinets. It felt particularly
comforting to escape to an area of research which would produce
results that had almost no chance of offending anybody.

When I broke for lunch, I went to meet my colleague Ian Tattersall
in his office in the Department of Anthropology. He was not there
at the moment, so I went into the lab across the hall and rummaged
through all the drawers that contained skulls of nonhuman hominoids.
The orang-utan had a small foramen lacerum. The others did not.
And the orang-utan still had a single incisive foramen while the African apes had their respective versions of the canals palatally. (You
always have to reconvince yourself that you saw what you saw.) Ian
came back, with Niles Eldredge in tow, and we went out for our
traditional lunch of reunion. Ian and Niles's book *The Myths of
Human Evolution* had just come out, Niles was working on the au-
tobiographical account of his role in the development of the model
of evolution called "punctuated equilibria," and I thought about humans and orang-utans.

Upon my return to Pittsburgh, I went on with my survey of hominoids. I had reached the limit of what information I could strain
from Schultz's and others' work that related to teeth, skulls, and
bones. Out of curiosity about what other areas of investigation might
contribute to this review, I decided to leave my preliminary comparisons of the basicranium until a later date.

*

Aside from teeth and bones, there are, of course, other parts of an animal, soft parts, that contribute to making it a whole, functioning organism. But these soft anatomical features have not usually been looked at with an eye toward phylogenetic reconstruction, especially by those who deal primarily with teeth and bones. Adolph Schultz, however, was not so limited.

Sitting on a wooden bench, a sculpted bentwood chair, or even a chair with a molded plastic seat is not something I can do comfortably for any length of time. I have had to sit on rocky or gravelly surfaces for hours on end while excavating a burial or chipping away at volcanic rock to expose the fossilized remains of tiny Eocene mammals and lizards captured within. But I don't like it — the sitting, that is. I like it even less if the surface is hot. A thick pair of denims just doesn't give me the protection needed. Basically, after a while, it hurts.

If I was a baboon or any other Old World monkey, I would have an easier time of it. I would have two thick, toughened patches of skin, one on each buttock. I would be born with these patches already growing in the right places, just as an ostrich is born with its heavy knee callosities already under way, or a newborn knuckle-walking African ape has its finger friction pads protecting its unused joints. I wouldn't have to wait until the irritation of constant sitting around on hot, rough surfaces caused these callosities — called ischial callosities because they are situated over the bone of the pelvis called the ischium — to form. A full-blown ischial callosity is large and hairless and has a thick, horny (keratinized) outer layer of skin.

As a group, the extant hominoids are distinguished from the Old World monkeys by the retarded development of ischial callosities in

FIGURE 7H   Buttocks of *(from left to right)* a baboon, a chimpanzee, an orang-utan, and a human, illustrating the extreme development (in the baboon) or complete absence (in the human) of thickened, "horny" ischial callosities. The orang-utan, while on occasion having patches of furrowed skin on each buttock, is most like humans in not having true callosities.

the hominoids. Gibbons do develop ischial callosities, but these structures appear later in the animal's life and do not get to be very large. Among the African apes, some individuals do not show any sign of ischial callosity development, but many do. Schultz and others studying this region found that ischial callosities are more frequent in the chimpanzee than in the gorilla. However, regardless of their frequency or size, ischial callosities in chimpanzees and gorillas are truly ischial callosities, like those of a baboon, complete in their "horniness."

Orang-utans rarely develop ischial callosities, and when they do, these callosities are weakly developed and not keratinized. Humans are unique among catarrhines in their total lack of ischial callosity development. To quote Schultz, "It is interesting to find that in regard to the frequency and development of ischial callosities man and chimpanzee stand much farther apart than man and orang-utan."

Another striking feature of humans and orang-utans is that they, among the large hominoids, develop the most widely separated nipples and mammary glands. In monkeys, the nipples are fairly close to the midline of the chest. They are farther apart in the chimpanzee and the gorilla. They are even farther apart in humans and are the farthest apart in orang-utans. Schultz commented on how humanlike the position of the mammary glands of an orang-utan is; at times they are quite close to the axial region (armpit). In having the most widely emplaced mammary glands, the orang-utan emerges yet again as being the most different of the large hominoids. However, once again, this difference is not one of primitiveness but one of additional anatomical novelty.

Aside from his astute observations, Schultz made some fairly insightful guesses as well. For example, he suspected, from what little he could observe and from the poor records at the time, that orang-utans and humans have the longest pregnancies — gestation periods — of any hominoid, and certainly of any primate. According to the latest information collected by Bob Martin, Susan Kingsley, and others in England studying the reproductive physiology of primates, Schultz was correct and, even more interesting, humans and orang-utans may have gestation periods of the same number of days. Chimpanzees seem to have, on average, the shortest gestation period of any of the large hominoids: a chimpanzee typically carries her fetus for about two hundred forty-five days. A gorilla usually gives birth after about two hundred sixty days. Humans, for which there is the most data available, have gestation periods that last about two hundred seventy days — and that appears to be the length of the gestation period of the orang-utan as well.

What makes this information even more exciting is that there does not appear to be a correlation between a hominoid's size and the length of its gestation period. One would think that the larger the hominoid, the longer its gestation period. And indeed, the chimpanzee, which is the smallest of the large hominoids, has the shortest average gestation period. However, the gorilla, the largest of the large hominoids, has a shorter gestation period than the somewhat smaller human and the definitely more diminutive orang-utan.

Along with this good guess on lengths of gestation, Schultz suspected that female orang-utans differ from female chimpanzees and gorillas in a way that human females do — by not developing any swelling of the genital region during the menstrual cycle, particularly at the time during the menstrual cycle when ovulation occurs. As I mentioned in chapter 1, most female animals come into estrus, or "heat," during the phase of their menstrual cycle when they are most likely to become impregnated, near ovulation. It makes sense that if there were going to be any physical or physiological cues to mate, they would occur at this time.

Although physiologists are discovering the increasingly important role of chemical stimulants, broadly referred to as pheromones, during the estrous cycle, it has long been recognized that there are visual cues associated with estrus. In most cases, the female's genital region swells for a time equal to or longer than the number of days it takes her to ovulate. In some species, such as the chimpanzee, this swelling is extremely dramatic and persists for a longer period of time (before and after) than does the phase of ovulation. In other animals, such as the gorilla, genital swelling is minimal and does not precede or extend beyond the ovulatory period by many days. There are even some mammals in which the male responds to the female cues for mating, not just by mating and paying some additional attention to her, but with a visible physiological response as well. In the mandrill, for example, a species of Old World monkey, the male's usually bright blue and red elongate snout becomes even more fiercely colored as the female's swelling and coloration peak.

Ron Nadler and others at the Yerkes Regional Primate Center in Atlanta, Georgia, and Charles Graham, formerly of Yerkes, have been studying the physical correlates of estrus. They have discovered that in contrast to the African apes, the orang-utan does not develop any genital swelling during her menstrual cycle. In this regard, the orang-utan female is again uniquely similar to the human female, and both are unique among the large hominoids in lacking an estrous cycle.

But there is more to estrus than just a visual component. An estrous cycle imposes limits upon mating. Cats, dogs, horses, and lemurs, for example, cannot have intercourse except during the estrous phase of the menstrual cycle. The same is true of chimpanzees and gorillas, but not of humans and orang-utans, who do not have estrous cycles.

A human female is potentially sexually receptive throughout her menstrual cycle. Intercourse is possible at any time. Studies have shown that human females tend to have intercourse more frequently around the time of ovulation than at other times during the menstrual cycle, but the possibility of spontaneous intercourse — at any time — is there. That characteristic was once thought to be distinctly human, unique among animals. But Ron Nadler, Charlie Graham, and their colleagues have observed the sexual behavior of captive orang-utans and found that orang-utans, like humans, have intercourse freely and at any time of the month.

In the wild, orang-utans are not very social, and great distances usually separate individuals. But there have been enough people observing orang-utans for long enough now to piece together a picture of orang-utan mating behavior. It is clear that orang-utans are not like the African apes.

First of all, when a female orang-utan is ready to mate (they do so quite infrequently, there being up to seven years between offspring), she forms a partnership with an adult male. And this partnership is for many weeks, not just a brief period around the time of ovulation. Second, the female orang-utan plays an active role in soliciting sexual advances from the male. Bouts of "lovemaking" are quite long — no quick thrusts of intromission as characterize most mammals, including chimpanzees. And third, it now appears that the same orang-utan male pairs up with the female when she enters the next birthing phase of her life. In short, the essentials of what is supposed to make humans sexually unique are shared with — and shared *only* with — orang-utans.

*

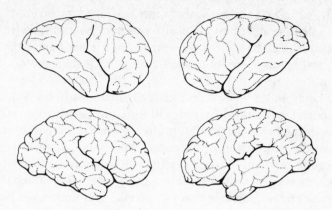

FIGURE 71 The right *(left)* and left *(right)* cerebral hemispheres of the brain of an orang-utan *(top)* and human *(bottom)*. Although they differ in size, the cerebral hemispheres of orang-utans and humans are unique among primates in being asymmetric in the expansion of the right frontal and left posterior (occipital) poles as well as in the deflection of the right Sylvian sulcus (solid line), which becomes quite vertically oriented.

Most mammals display some asymmetry between the right and left sides of the brain in size and morphology. Among mammals, humans are always cited as having the most extreme development of cerebral asymmetries. It is usually the right frontal lobe of the cerebrum — that "thinking" part of the brain which has become grossly enlarged in humans and dolphins — that is larger than the left frontal lobe. At the posterior end of the cerebrum, the left occipital lobe is usually larger than the right occipital lobe; in some individuals, however, these asymmetries are reversed. (There are some others in whom no measurable cerebral asymmetry ever develops.) Cerebral lobe asymmetries in humans are supposed to be correlated with the development of hand use preference, that is, for right- or left-handedness.

Humans have also been noted for having the greatest asymmetry in length and deflection of the Sylvian sulcus, a groove in the surface of the side of the cerebrum, just above the temporal lobe. On the left side of the cerebrum, the Sylvian sulcus originates from the area of contact between the frontal and temporal lobes and courses posteriorly at a low, almost horizontal angle. The right Sylvian sulcus, however, is not only longer than the left one, it does not run such a relatively straight course. Rather, the last third or so of the right Sylvian sulcus is deflected vertically so that, instead of terminating posteriorly along the side of the cerebrum, it ends nearer the top of the cerebral hemisphere, not much farther back than the cerebrum's midpoint.

Neuroanatomists believe that such asymmetries in the Sylvian sulcus are correlated with differences in function between the right and left sides of the brain. Right–left brain asymmetries are also probably related to the acquisition of language. And humans, with their typically large brains and marked cerebral and sulcal asymmetries, have language, as well as a preference for using one hand rather than the other. Even microcephalic humans — people with undersized brains — still spontaneously develop language and handedness.

In a series of studies on right–left brain asymmetries in primates, Marjorie LeMay, Norman Geschwind, and their colleagues at Harvard Medical School made some interesting discoveries. One was that although all primates display some degree of right–left brain asymmetry, the greatest degree of asymmetry occurs in the large hominoids. Ralph Holloway, of Columbia University, who has struggled for years to perfect the analysis of brain casts taken from the inside of the skulls of fossil hominids, subsequently found that the gorilla had an unexpectedly marked asymmetry in the size of the occipital lobes of the cerebrum. But overall, and especially in the deflection of the right Sylvian sulcus, LeMay and Geschwind found that next to humans, the orang-utan had the greatest amount of right–left cerebral asymmetry, the chimpanzee noticeably less, and the gorilla the least. The understated abstract at the beginning of one of LeMay and Geschwind's articles says it all: "Asymmetries are found in the cerebral hemispheres of some great apes, particularly in the orangutan, that are similar to those seen in man. Studies in the orangutan might be more likely to help in understanding the evolution of handedness or language than studies in chimpanzees."

Of all the novelties I had come across which humans and orang-utans share, and seem to share so exclusively, I was most surprised to find such shared uniquenesses in their brains.

The whole matter of brain asymmetries and function — and the possibility of something evolutionarily important being there to investigate — became even more tantalizing when I happened upon a recent article in the *Journal of Human Evolution* entitled "Handed-ness in *Pongo pygmaeus* and *Pan troglodytes*." The authors of this study, Brésard and Bresson, had seen LeMay and Geschwind's work and commented on the apparent contradiction that the orang-utan, which "seems to belong in evolution to an earlier branch than man," has "the strongest morphological likeness to humans" in brain asym-metries.

What did Brésard and Bresson find? The orang-utan in their study was consistently and spontaneously right-handed in tests that required only the use of one hand to perform the task as well as in those tests where both hands were needed to achieve the objective. The chim-panzee, however, alternated between its left and right hand as the preferred hand to use during the unimanual and bimanual tests, al-though when knuckle walking it did seem to favor the left hand.

This must be the tip of the iceberg. I have been told by some who work with the large apes in captive situations, such as teaching them sign language, that the orang-utans are the ones which are noticeably handed in signing as well as in performing other tasks. But so far, long-term studies of handedness in the apes have not been published. Studies of this kind are especially important because if there is going to be a measurable expression of behavior correlated with right–left brain asymmetries, it is going to be in the study of handedness. All the apes can be taught to communicate in sign language.

The significance of human and orang-utan right–left brain asym-metries might also be reflected in a different arena of their anato-mies — in particular, in aspects of their reproductive physiology. As mentioned in the preceding chapter, the large hominoids are set apart from other catarrhines in that the females excrete estriol as the major component of their estrogen. The major urinary estrogen excretion of a female monkey, for instance, is a steroid called androsterone. The large hominoids have a noticeably high level of estriol excretion during the normal menstrual cycle as well as during pregnancy. During pregnancy, estriol excretion is correlated with adrenal gland devel-opment and steroid secretion of the fetus. As Bill Lasley, Nancy Czek-ala, Susan Shideler, and others have suggested, it appears that fetal adrenal growth and function are correlated with normal brain growth of the fetus.

The story of the estriols does not stop with the large hominoids, with their last common ancestor. Lasley, Czekala, and Shideler also discovered that humans and orang-utans have the highest levels of estriol excretion of the four large hominoids. This noticeably higher level of estriol excretion occurs not just during the normal menstrual cycle, but during pregnancy. In fact, estriol levels in humans and orang-utans are an incredible four to five times higher than in chimpanzees and gorillas.

This discovery was so dramatic that Lasley and his co-workers concluded that humans and orang-utans must be different from the chimpanzee and the gorilla not just at the placental level of mother–fetus interaction, but in fetal development altogether. They suggested that among the large hominoids there are two types of fetal–placental interactive units. One type is found in the African apes. The other distinguishes humans and the orang-utan.

Since maternal estriol levels during pregnancy are reflective of adrenal function and activity of the fetus, and since fetal adrenal function and activity are apparently correlated with proper fetal brain growth, it may not be mere coincidence that estriol levels during pregnancy are so extremely high in humans and orang-utans. After all, these two hominoids also develop the most marked cerebral asymmetries.

At this point in my review of the literature, I had not gleaned much to support any of the more popular schemes of relationships among the large hominoids. To my surprise, I could not piece together a robust set of homologies indicating overwhelmingly that humans were closely related to one or both of the African apes. Instead, once the snowball got rolling, the association of humans with the orang-utan became the most frequently corroborated and compelling evolutionary proposition. From all aspects of hard- and soft-tissue anatomy, as well as reproductive physiology and development, there emerged a set of apparent homologies the likes of which I have rarely encountered when trying to unravel the phylogenetic relationships of other animals.

It seemed inescapable. Our closest living relative is the orang-utan.

What I found particularly amazing about my review of the literature was that so many different aspects of these hominoids had been sampled — hair, teeth, areas of the skull, mammary glands, gestation periods, sex hormone levels, the brain, arm development, and even sexual behavior — that for the first time in all the years I have been attempting to reconstruct the evolutionary relationships of organisms, I had a real and tangible sense of a common ancestor as a total, living, breathing, thinking, behaving, reproducing animal, not just as a vague theoretical construct.

I worked up a lengthy paper based on these findings. To put it into perspective, I began with a review of the history of thought on who constituted our closest living relative or relatives. I also incorporated the fossil material into my discussion and offered what I thought was a critical discussion of the underlying notions used in molecular approaches to phylogenetic reconstruction. Since the British journal *Nature* had a history of publishing articles on newly discovered fossil hominoids, and had more recently featured articles by Peter Andrews as well as David Pilbeam on the reinterpretation of the evolutionary relationships among the hominoids of *Sivapithecus*, I thought I would make an inquiry about their publishing my conclusions.

After an unusually long period of waiting for a response (I have had papers rejected by *Nature* in record time), the editors of *Nature* wrote back that they would consider my manuscript for a review article (which is an outlet usually reserved for the more senior in the profession) if I were to reduce its length — a lot.

Thank goodness for word processors. After copying the manuscript file, I was editing with a vengeance. Fifty pages became eighteen. It was an educational experience in brevity.

The new manuscript went back to England and, after receiving treatment from three outside reviewers, was accepted. It was to be published to coincide with the opening of the "Ancestors" exhibit at the American Museum of Natural History, which was the first major exhibit of representative specimens of all species of *Australopithecus* and *Homo* — together, in one place. To be at the "Ancestors" gathering, to study some of the world's rarest hominid specimens, and to have a review paper coming out at the same time was unbelievable. I even ended up giving a presentation of my human–orang-utan theory during the paper sessions at "Ancestors." I had had it prepared to give at the meetings of the American Association of Physical Anthropology the following week, so I filled in for colleagues who couldn't attend "Ancestors."

I was petrified. But I was told it went all right, that I had been articulate and well organized. And that I had gotten some members of the audience to take notice.

I floated to Philadelphia for the next round of meetings knowing that my presentation was a good one — or at least well organized.

Upon arriving at the hotel hosting the gathering, I went to register and bumped into Matt Cartmill, my long-term sparring partner on the evolution of the tarsier, who promptly commented on the photograph of the skulls of the four large hominoids that were peering out from the cover of that week's *Nature*. I had even gotten the cover. I had sent in a few slides of different views of these skulls to be considered for use on the cover, but no one had mentioned the matter after that. I couldn't believe it.

But this sudden elevation to worldwide visibility was accompanied by greater vulnerability. It also meant that I had an even greater obligation to poke at my theory, to try with increased diligence to find out where my theory might fail, and to investigate why it differed so much from what everyone else seemed to take as the truth. In short, I now had to convince myself that my theory should be discarded in favor of the results derived from the study of molecules and chromosomes.

8

Blood's Thicker
Than Bones

"THE GENE'S where it's at," the biochemist kept saying during two back-to-back, exhausting meetings.

I probably wouldn't have phrased it quite that way, but the general sentiment reflects a growing point of view. On some level, it seems intuitively appealing that the closer one gets to the genetic level of an organism, the closer one is to all sorts of answers, especially those of relatedness. We even have a popular saying — "Blood's thicker than water" — to reflect an almost innate belief that there is something that binds the closely related together.

Sociobiologists have formalized a part of this sentiment in their approaches to evolutionary problems. One point of view held by sociobiologists is that if the body is the gene's way of getting itself from one generation to the next, then there must be something informative about looking at the number of copies of itself a gene leaves behind. In organisms such as mammals, an offspring receives half of its total complement of genes from each parent. Thus, a close relationship exists between parent and offspring by virtue of their sharing fifty percent of the parent's total genetic reservoir.

Sometimes an adult cannot (or, for various reasons, will not) mate, but he or she will still try to ensure the success of his or her genes in subsequent generations. For example, a mother will pass on fifty percent of her genes to her offspring, but half of these genes she holds in common with her own siblings, her brothers and sisters. Because of this sibling relationship, she will also be passing on to her offspring one half of the fifty percent of the genes she shares with her brother.

So even if a male does not himself produce offspring, one quarter of his genes will get into another generation anyway, passed on by his sister. It is to his benefit to protect his sister and especially her offspring, because his nieces and nephews hold twenty-five percent of his genes. In various human cultures, the mother's brother — the uncle — does indeed play an important role in the rearing of his sister's children.

An extreme situation is found in identical twins. Identical twins arise when a fertilized egg splits in half, which results in two identical "eggs." The two individuals that develop thus share identical genetic reservoirs, or genomes. The offspring of an identical twin, as would any other offspring, will inherit fifty percent of its parent's genes. However, because of this unpredictable embryonic phenomenon, the offspring of an identical twin will also be receiving fifty percent of its parent's sibling's genes — not twenty-five percent, as in the more usual cases of inheritance.

All of this goes to demonstrate what might appear to be obvious: the more closely related individuals are, the more of their genome they have in common. The corollary is that the more distantly related individuals are, the less genetic material they have in common. Our own notions of relatedness, observed through our own familial genealogies, seem to bear this out.

We can extend this notion even farther. If we think about our own kinship and extended families, we can envision increasingly larger circles of individuals who are still more closely related to one another than they are to other individuals. But there must be a barrier at some point, a limit to the strings of relatedness.

There is. It is what people generally think of as a species.

All individuals of the same species must be genetically more similar to one another than any of them is to individuals from any other species. If we think of species as individuals, we can justifiably entertain the idea that those species which are most closely related to one another are also the most genetically similar to one another. There is nothing, at least on the face of it, that seems to preclude there being some sort of equivalence between genetic closeness and closeness of relatedness. This certainly seems to be the case when these notions are considered at the level of the individual. And if it seems to work at one level, surely it must also be true more broadly.

In 1904, George H. F. Nuttall, University Lecturer in bacteriology

and preventive medicine at Cambridge University, published a monograph that would influence the study of evolutionary relationships
forever. It was entitled *Blood Immunity and Blood Relationship: A
Demonstration of Certain Blood-Relationships Amongst Animals by
Means of the Precipitin Test for Blood.* His goal was to demonstrate
the evolutionary relationships of animals by studying the degrees of
immunological or antibody–antigen reactivity which would be produced by combining blood serum with antiserum.

Many of us, without actually knowing it, are aware of various
examples of reactions involving the immune system. In general, most
people know that there are "things" in our blood system which somehow combat infection and disease. We are perhaps even more aware
of these "things" as a result of the AIDS crisis. We know that the
AIDS virus knocks out those "things" in our immune system, making
the AIDS victim unable to defend himself against the onslaught of
any disease.

An example of a type of immune response comes from study of
the blood's Rh factor (called Rh because it was first identified during
studies on rhesus monkeys). There are two forms in which an Rh

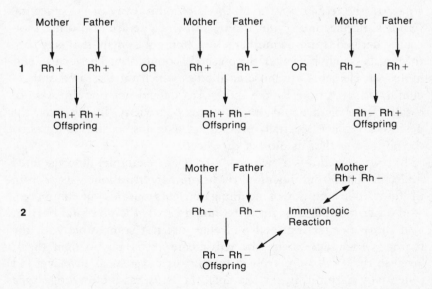

FIGURE 8A A simplified illustration of inheritance, using the Rh factor as an
example.

gene comes. It is either an Rh+ (Rh positive) or an Rh− (Rh negative). The Rh+ is dominant over — that is, its activity effectively masks — the Rh− gene. Since a child receives half of its genetic complement from each parent, it will inherit one Rh gene from its mother and the other from its father.

There are three ways in which an individual can be Rh+. One is by receiving an Rh+ gene from each parent. Another is by receiving an Rh+ gene from the mother and an Rh− gene from the father; since the Rh+ gene will dominate over the Rh− gene, the Rh+ gene will be the one expressed. The third way is by inheriting an Rh− gene from the mother and an Rh+ gene from the father. So there is an excellent chance that you will be Rh+, even if one of your parents carries an Rh− gene. But if both parents carry an Rh− gene (even if it's masked), there is a twenty-five percent chance that you will inherit it from each parent and end up as Rh−. And if your mother is Rh+, you — as a developing fetus with Rh− blood — are in trouble.

Although the fetus's blood never mingles with the mother's, the placental contact between mother and fetus allows molecules to cross between the two individuals. One of these molecules is oxygen, which comes to the fetus from the mother's blood by way of the placenta. Other molecules are of nutritional value. There are also other molecules, derived from the mother's own immune system, that seep into the fetus's system by way of the placental connection. These particular molecular elements are those antibodies which were created and accumulated over time by the mother to defend her body in case of assault from disease and other sources of toxicity. In this way, the newborn is at least partially pre-protected against various illnesses by the antibodies that its mother produced.

But the placenta is a two-way street. For example, deoxygenated molecules pass from the infant to the mother. Metabolic wastes from the fetus's consumption of incoming nutrients must also be eliminated. And other molecules might just wander across the placental barrier.

If these meandering fetal molecules are not compatible with the mother's immune system, she will create antibodies to fight them, because they will be perceived by her own system as invasive and potentially harmful. If, for example, the fetus has the "wrong" type of Rh factor, the mother will produce antibodies to them, to protect herself from these intruders. But the mother's antibodies will not just lie in wait for future waves of invasion; rather, they will cross the placenta and attack the "wrong" Rh antigen of the fetus.

Usually, the first fetus has a pretty good chance of survival. But if there is a second Rh− fetus in another pregnancy, the antibodies generated by the mother will be more concentrated and even stronger. Since the Rh factor plays a role in the proper functioning of the red blood cells (erythrocytes), sufficient oxygenation of the fetus's blood can be impaired. Total blood transfusions at birth may not even be sufficient to save the newborn. Fortunately, these days, genetic counseling and sophisticated techniques of fetal monitoring are available and greatly reduce the possibility of Rh-factor deaths.

Essentially, the antibody that the mother's immune system created can be thought of as an antitoxin. This antitoxin is the antidote produced to combat the antigen of the different Rh gene of the infant, which acts as a toxin. Various vaccines that we are given to protect us from serious illness — diphtheria, tetanus — are antitoxins that are the antibodies produced when an experimental animal is injected with the disease. If a scientist decided to inject me with a syringeful of a cat's blood serum, I would produce an antiserum to the proteins in that cat's blood serum. But one has to be careful about pursuing this type of project. A late nineteenth-century mad French scientist found that an injection of ten cubic centimeters of human blood was toxic enough to kill a rabbit.

It was during the late nineteenth century that scientists discovered how to produce an antiserum that would be highly toxic to the animal against whose blood the antiserum was made. For example, blood from a rabbit, injected into a horse, will cause the horse to develop antiserum to the rabbit's blood. This antiserum, when put back into a rabbit, will attack the rabbit's own blood serum. This happens because there is specificity between an antigen (in this case, blood serum proteins) and the antibody (the antiserum) produced. And there will be elements of an animal's blood serum that will be specific to that organism.

At the turn of the century, Nuttall and a German, Uhlenhuth, independently came to the conclusion that there might be something "worth while to study[ing] the reaction given by the bloods of *related* species." The task was to take a sample of blood from one species and introduce it into another animal, which would then produce an antiserum to it. One then takes this antiserum and sees how slight or

Testing "Blood Relationship"

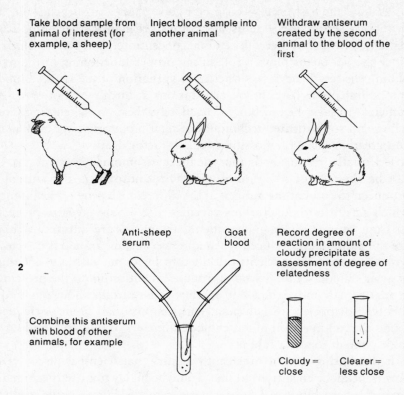

Take blood sample from animal of interest (for example, a sheep)

Inject blood sample into another animal

Withdraw antiserum created by the second animal to the blood of the first

Anti-sheep serum

Goat blood

Record degree of reaction in amount of cloudy precipitate as assessment of degree of relatedness

Combine this antiserum with blood of other animals, for example

Cloudy = close

Clearer = less close

FIGURE 8B A schematic representation of the basic procedure in testing "blood relationship" on the basis of immunologic reaction: the more similar the blood serum proteins of organisms, the greater their immunologic incompatibility.

severe the reaction is between it and the serum of other animals. The reaction throws off a precipitate, which makes the solution cloudy. The stronger the reaction, the cloudier the solution. The reaction is the most severe when the serum of an animal is combined with an antiserum made to the same animal.

Nuttall and Uhlenhuth predicted that the most closely related species would produce the most reactivity when the antiserum to one was combined with the blood serum of another. There would be more reactivity because a closely related species would have more sites in its serum to be attacked by the antiserum produced to its close relative. And indeed, as Nuttall noted, "Uhlenhuth found anti-sheep serum to precipitate sheep, goat and ox blood, anti-horse to precipitate horse and donkey blood, anti-human to precipitate human and monkey blood."

The strength of the antiserum played an important role in determining just how many species might belong to the same group. A weak antiserum would provoke reactivity in the sera of only the most closely related of organisms, while a stronger solution would produce a precipitate when combined with the sera of a greater number of related animals. There would be a proportionate decrease in the degree of reactivity between serum and antiserum as one got to the more distant relatives in the group. In addition to strength of reactivity, Nuttall felt that the rate at which a reaction took place was also an indicator of the degree of relatedness.

Their experiments culminated in 1901, with the proposition that "the zoological relationships between animals are best demonstrated by means of powerful antisera." Essentially, a "blood relationship" was a zoological and, by extension, an evolutionary relationship. As Nuttall wrote, "if we accept *the degree of blood reaction as an index of the degree of blood-relationship* within the Anthropoidea, then we find that the Old World apes are more closely allied to man than are the New World apes, and this is exactly in accordance with the opinion expressed by Darwin."

True, this was Darwin's opinion. It was in large part derived from Huxley's *Man's Place in Nature*. Bearing in mind that "New World ape" refers to the New World monkeys, whereas "Old World ape" includes the Old World monkeys as well as the true apes, Nuttall's immunologically derived conclusions agreed in rough outline with a morphological arrangement of the major groups of primates.

In his 1904 major monograph Nuttall brought his immunological approach to "blood relationships" to bear on all orders and classes of animals. He sampled and tested serum—antiserum reactions on representatives of most mammals, various birds, reptiles and amphibians, and even a few crustaceans, including a lobster. It turned out the general arrangement of "distances" between classes of organisms, and the general groupings within classes of organisms, coincided with the general arrangement of animals derived through the more traditional means. Nuttall even demonstrated the greater "blood relationship" between humans and the large apes (the gibbon was not included in the study) than between humans and the Old World monkeys. This, too, was consistent with Huxley's human–"great ape" notion of large hominoid relationships.

Nuttall's impact on the evolutionary world was not to be immediate. Decades passed, during which time population genetics became increasingly sophisticated and molecular genetics rose to predominance, largely through the discovery by Crick, Watson, and others of the molecules basic to life and inheritance: DNA (deoxyhydronucleic acid), RNA (ribonucleic acid), and their collateral molecular kin (for example, mRNA, messenger RNA).

The early 1960s saw a minor revolution in the evolutionary sciences, when Nuttall's notion of "blood relationships" was resurrected with a vengeance. Not only were serum–antiserum reactions studied throughout most of the animal kingdom, but the actual elements of blood serum were being identified and compared among organisms.

A major figure in the development of molecular and biochemical approaches to systematics — which, when applied specifically to primates and matters of human evolution, is referred to as molecular anthropology — is Morris Goodman of Wayne State University in Detroit. More than almost anyone else, Goodman was responsible for demonstrating the acceptability of using immunochemical studies of blood serum proteins (and, as a consequence, any molecule) for unraveling the evolutionary relationships of organisms. Goodman refined and augmented Nuttall's work in the demonstration of the unity of Anthropoidea; the separateness within Anthropoidea of the New World monkeys (Platyrrhini) and Old World monkeys and hominoids (Catarrhini); the separateness within Catarrhini of Old World monkeys and the hominoids (Goodman did include the gibbon in his studies); and, within Hominoidea, the apparent unity of humans and the two African apes, the chimpanzee and the gorilla, to the exclusion of the gibbon and the orang-utan.

As early as 1962, Goodman was arguing that humans and the African apes were so similar immunologically that they should be placed together in the family Hominidae. But it was this leap into the realm of taxonomy and classification that, in retrospect, seems to have been one of the major reasons molecular anthropologists had to fight so hard to get their ideas accepted by the traditional, morphologically oriented evolutionary biologists. Genealogy or evolutionary relatedness is one thing. The pursuit of similarity in such matters was certainly acceptable practice. Classification was another thing altogether. No proper systematist would want to call a chimpanzee or a gorilla a hominid.

Other obstacles to the acceptance of biochemically derived conclusions about relatedness revolved around the details of various suggested relationships. On the more general levels of phylogeny — for instance, the existence of a group such as Anthropoidea, and its membership broadly of monkeys and hominoids — there was not a whole lot of disagreement between molecular systematists and paleontologists and morphologists. However, there usually was some major contradiction between the phylogenies derived from these different approaches in terms of the exact evolutionary positions of specific taxa — for example, the tree shrews or the tarsiers. As long as there remained discrepancy between the results of molecular systematists and those of morphologists, and especially among molecular systematists themselves, there really was no compelling reason for leaving the comfort of the more traditional morphological and paleontological approaches to primate systematics.

In order to study the gross differences and similarities among primates in their blood serum proteins, Goodman devised a simple procedure to produce the necessary reactions between serum and antiserum. Serum from an animal was placed on a small plate, the required antiserum was placed on an adjoining plate, and a third plate received both serum and antiserum as they diffused from their plates of origin. Once on the common third plate, the sample of serum would interact with the sample of antiserum. The strongest reactions — between the serum and antiserum with the most sites in common — produced a very visible thick band that arced across the common plate. Reactions between serum and antiserum with increasingly less comparability produced weaker bands or bands that did not arc completely. In these instances, it was common to see the band continuing as a spur that took off tangentially from the semi-arc and proceeded into the farther reaches of the plate, where it faded out.

While Nuttall and Uhlenhuth had to be content with studying the immunological reactions among animals in their whole blood sera and antisera, subsequent advances in biochemistry made it possible for Goodman and fellow molecular systematists to study the different molecules that together constitute an organism's blood serum. With regard to two of these blood serum components, ceruloplasmin and transferrin, Goodman found that "gorilla and chimpanzee appeared to be identical with man, whereas gibbon and orangutan diverged from man." However, study of another of the blood serum molecules, albumin, yielded a slightly different picture. In terms of albumin reactions, it appeared that the gorilla was most similar to humans. The chimpanzee was the next most similar to humans, after which came the gibbon. Least similar to humans in its albumin was the orangutan, which emerged as the most divergent — or different — of the hominoids.

When Goodman looked at the blood serum protein macroglobulin, he found that humans and the gorilla were the most similar, as they seemed to be in their albumins. However, an analysis of gamma globulin demonstrated the greatest similarity between humans and chimpanzees, with decreasing similarity between humans and the gorilla, the orang-utan, and the gibbon.

The proteins of the blood serum can also be studied by a process called electrophoresis. This is accomplished by placing a sample in a well at the base of a starch-gel column through which an electrical current is passed. Since the surface of a molecule has an electrical charge, an electrical current excites the molecule to move, proceeding up the column. All of the molecules present in the sample at the base of the column are charged up to scale the column, but not all molecules will make it to the top. Different molecules travel through the starch-gel at different rates.

The speed at which a molecule or protein proceeds up the column is determined by the molecule's size and weight. The smaller and lighter molecules will end up higher in the column than will the heavier and larger ones. Different molecules will also leave specific "fingerprints" in the shape and density of the blob or band they produce on the starch-gel. The final picture — the portrait of blood serum proteins specific to any given animal — is seen as a series of blobs and bands of different widths, shapes, and thicknesses that are spread out in a recognizable pattern throughout the starch-gel template.

Goodman used electrophoresis to study the similarities and differences among the hominoids in their blood serum proteins. He found that the "hominoid patterns show striking divergences from each other," thus arguing, he felt, against "placing chimpanzee and gorilla in the same genus," a proposal that has enjoyed occasional popularity.

But perhaps more important, Goodman concluded that "as judged by the overall impression of the patterns, man shows more similarity to the gorilla and chimpanzee than to the gibbon or orangutan." The gibbon and the orang-utan appeared to be the most divergent from humans.

My impression of the starch-gel patterns is in partial agreement with Goodman's. Indeed, the gibbon and the orang-utan do appear to differ slightly from the other hominoids. But it may not be the case that this difference is caused by those hominoids' primitiveness relative to the others. For example, the human pattern, and to some degree, the patterns of the African apes, seem to be quite like those of the various Old World monkeys Goodman analyzed — which would imply primitiveness for these hominoids and individual derivedness for the gibbon and the orang-utan. But within the general state of catarrhine primitiveness, the chimpanzee and gorilla seem to have their own peculiar banding of the transferrin component of their blood serum.

Obviously, the criterion of overall similarity was firmly established in the early phases of molecular and biochemical analyses aimed at determining relatedness. In the case of trying to figure out to whom humans were most closely related, the approach was one of trying to find the best match among the hominoids to the focal organisms, humans. Thus, for example, in electrophoretic patterns of blood serum proteins, humans and the African apes emerge as a better match than humans and orang-utan or humans and gibbon — even though the broader comparison, which was available, illustrates that the ultimate match to humans is found among the Old World monkeys.

Another way in which humans were linked to the African apes was along the lines of the approach used by William K. Gregory. Goodman concluded that his serum–antiserum analyses "show that the Asiatic apes are more distant from man than are the African apes." However, some molecules (albumin, for instance) were most similar in humans and the gorilla, while other molecules (gamma globulin, for example) were most similar in humans and the chimpanzee. It seems to me that even if we assume this information does translate directly into phylogeny, it does not add up to humans' being most closely related to both of the African apes.

In 1966 Vincent M. Sarich and Allan C. Wilson published in *Science* the first of three papers that were to shake the morphological and paleontological world further. Sarich had been a graduate student working in Wilson's lab at the University of California, Berkeley. Sarich's dissertation project was to investigate the evolutionary relationships among primates using biochemical approaches, and Wilson had established in his lab a technique developed by Wasserman and Levine that yielded great immunological reactions using minute amounts of serum and antiserum. From a purely practical point of view, this technique was a real improvement. The technique was called micro-complement fixation (abbreviated as MC'F), and the title of Sarich and Wilson's paper was "Quantitative Immunochemistry and the Evolution of Primate Albumins: Micro-Complement Fixation."

Albumin is a large molecule and, compared to other blood serum proteins, relatively easy to separate and purify. Sarich and Wilson tested first serum–antiserum reactions among humans, the chimpanzee, the rhesus monkey (to represent the Old World monkeys), and the capuchin monkeys (the organ grinder's monkey, to represent the New World monkeys). Reactivity was greatest between human antiserum (actually, antihuman serum) and chimpanzee serum, and least between antihuman serum and the serum of the capuchin monkey. From these observations, Sarich and Wilson concluded that "these species form an approximate evolutionary series, the chimpanzee being the closest relative of man and the capuchin monkey the most distant."

By using increasingly stronger concentrations of antiserum, eventually reaching a peak of reactivity between serum and antiserum, Sarich and Wilson could calculate the "index of dissimilarity," which subsequently became known as the "immunological distance" or simply "I.D." The lower the I.D. value, the more similar the animals under study were supposed to be in their albumin, and thus the more closely related. Higher I.D. values were supposed to reflect greater dissimilarity between the organisms being analyzed, and thus greater evolutionary distances. Perfect identity — which is what one would get by testing antihuman serum against human serum or antigibbon serum against gibbon serum — is 1.0. The closer the I.D. value is to 1.0, the closer the molecular identity and the evolutionary relationship.

To check their initial determination of immunological distance, Sarich and Wilson devised the test of reciprocity. For example, and this is admittedly a simple example, if chimpanzee serum was cross-reacted with antihuman serum the first time around, human serum was cross-reacted with antichimpanzee serum the next time to see if the resultant I.D. values were reasonably similar. As one might have predicted, the test of reciprocity usually confirmed the initial immunological finding.

Since this, the first of the Sarich and Wilson papers, was primarily a test of the applicability and accuracy of the MC'F technique, Sarich and Wilson were pleased to find that their I.D.'s yielded an arrangement of primates in general agreement with the results of most other immunological studies. They did, however, disagree with the findings of some previous studies on albumin which concluded that the gibbons were about as dissimilar as Old World monkeys were to human albumin. To the contrary, Sarich and Wilson derived I.D. values that clearly placed the gibbon closer to the other hominoids than any Old World monkey, but not as close as the large apes. This result also differed from Goodman's analysis of albumin, in which the orangutan when compared to humans emerged as more dissimilar than did the gibbon. Outside of the hominoids, Sarich and Wilson found that they were also in some "slight disagreement in the relative placement of the various prosimians."

These occasional inconsistencies aside, it did nevertheless seem that the general arrangement of the primates arrived at by Sarich and Wilson was more consistent with the going world view than it was in disagreement. Because of this, Sarich and Wilson concluded that "the MC'F data are in qualitative agreement with the anatomical evidence, on the basis of which the apes, Old World monkeys, New World monkeys, prosimians, and nonprimates are placed in taxa which form a series of decreasing genetic relationship to man."

But here, as in other groundbreaking studies in molecular systematics, the source of validation of the approach and its results was a morphologically based phylogeny. Minor variations in the placement of specific taxa were allowed. If the general arrangement of groups was reasonably similar, then the molecular approach used was seen as viable. And, of course, as is true among all manner of systematists, overall similarity — here identified by I.D. values — was the overriding criterion for determining relatedness.

You would think that even though Sarich and Wilson's results were not in total agreement with Goodman's on the relationships among the primates and other mammals that would eventually be studied, these three molecular systematists would have formed a biochemical front to the morphological majority of systematists. But this seemingly natural alliance was not in the cards. Instead, the years that followed witnessed some rather bitter battles in the literature between these two camps. Goodman came to head a group out of Wayne State, and the Sarich and Wilson duo expanded to include Jack Cronin.

Some of the debate centered around the placement of single taxa — for example, the tarsier. Other energies were focused on the rates at which molecules change. This latter pursuit assumed a larger-than-life importance, not just among molecular systematists, but among paleontologists, because rates of molecular change were used to calculate the ages of lineages. Once one knew how long any given taxon had been around and could determine the times at which sister taxa diverged from their common ancestor, the last bastions of paleontology came under direct attack. Molecules, not fossils, were now supposed to guide us through the evolutionary history of organisms.

In the 1960s, *Ramapithecus* (known these days as *Sivapithecus*) was the acknowledged ancestor of the proper hominids, *Australopithecus* and *Homo*. Because *Ramapithecus* was known from volcanic deposits in East Africa, the age of the fossil could be chemically determined. It was dated to the Miocene, as old as or older than fourteen million years. The broader implication of associating *Ramapithecus* with later hominids was that the hominid lineage could claim a substantial chunk of evolutionary history for itself. A mid- or older Miocene date for *Ramapithecus* meant that hominids as a group could not have diverged from their closest hominoid relatives more recently than fourteen million years ago. Since, by the 1960s, the African apes were gaining popularity as the better choice for our closest living relatives, the scheme that became popular was that the African apes and hominids diverged at least fourteen, perhaps even eighteen, million years ago. This seemed to fit in with other components of the generally accepted picture of hominoid evolution, which tied the emergence of the large hominoids to the Miocene apelike fossils *Dryopithecus* and especially *Proconsul*, which had been found in deposits that were even older than those that yielded *Ramapithecus*.

Goodman interpreted the immunological and biochemical data as being compatible with this prolonged time scale of diversification among the larger hominoids. He invoked selectionist arguments to explain the apparent acceleration or deceleration in rates of molecular change that would have had to have occurred if one accepted the paleontologically established dates for the earliest appearances of each

of the hominoids. Thus, for instance, if the origins of the various
hominoid lineages had indeed occurred over ten million years ago,
the extreme similarity in the albumin of the various extant hominoids
had to have resulted from a slowdown in the rate of molecular change
of that particular blood serum protein. And Goodman had a very
good explanation for why albumin evolution would have slowed down
in the hominoids.

The reason, Goodman argued, for such deceleration of change in
the albumin molecule is related to the fact that the hominoids have
a hemochorial type of placentation and very prolonged gestation periods. Hemochorial placentation — the most intimate of placental
modes in the association of the maternal and fetal blood systems —
is common to all anthropoids. Hominoids, especially the large hominoids, have much longer gestation periods than other anthropoid
primates. In animals with fairly short gestation periods, there is enough
slack in the system to allow for unimpeded molecular change to occur —
the fetus will be born before the mother's immune system can produce
anything that might be lethal or injurious to the fetus. But a hominoid's gestation period is just too long for any serious maternal–fetal
incompatibility to exist. Selection would have to act to reduce the
possibility of immune responses from the mother toward her fetus.
Thus, there would have to be a slowdown of molecular change that
was commensurate with the period of gestation. Albumin was obviously one of the molecules that did just that.

Sarich and Wilson preferred a different explanation for the extreme
similarity in the hominoid albumins. In their third paper, "Immunological Time Scale for Hominid Evolution," published in 1967,
albumin similarity was taken at face value. Little difference in albumin — little molecular change — meant that little time had elapsed
since the separation of the hominoids. Since they had argued in their
second paper that molecular change in albumin was constant among
the major groups of primates, Sarich and Wilson converted the observed hominoid albumin differences into a time scale. Molecular
change clicked away at a constant rate, they claimed, and thus a
"molecular clock" could reveal the times at which the various hominoids — actually any species — diverged and went off on their own
evolutionary paths.

But if molecules did change at a constant rate, you had to calibrate
the clock somehow so you could tell how long the different animal

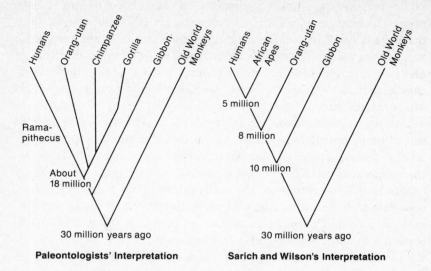

FIGURE 8C Traditional paleontology and morphology versus the "new" approach of molecules and the molecular clock: who's related to whom and how recently did evolutionary events occur?

lineages had been evolving — that is, you had to find a date, somewhere, by which to set your clock. Only then could you calculate the real rate at which a molecule's bits and pieces were changing every ten thousand or one million years. The only place you could get a date was from the fossil record. And here you had to assume that the fossil record had been interpreted properly.

Sarich and Wilson took the split between the Old World monkey lineage and the hominid lineage as their starting point. Most paleontologists were in agreement that the split occurred about thirty million years ago. Now Sarich and Wilson could calculate the date for the separation of the individual extant hominoids. And they did. I.D. values, originally taken as direct assessments of evolutionary distance, were converted into chronological milestones of evolutionary events.

The result of their calculations was that the gibbon had diverged about ten million years ago. The orang-utan had become distinct at eight million years. Finally, humans and the African apes had separated a mere five million years ago.

Blasphemy. That was essentially the response of virtually everyone, regardless of whether they were molecularly or morphologically disposed. Humans and the African apes, assuming that they were sister taxa, could not have become distinct only five million years ago. Much more time was needed for natural selection to bring about the obvious changes that had to have occurred to make humans so different from the chimpanzee and the gorilla. If hominids came into existence only five million years ago, then by definition *Ramapithecus* could not be a hominid, a near-hominid, or a hominid ancestor, or have anything to do with hominids. Even when Sarich later recalculated his molecular clock and added a few million years to the human lineage, *Ramapithecus* was still *hominid non grata*, an ancestor out in the cold.

Goodman, Sarich, and many others argued bitterly over the "constancy" of the molecular clock. As Simons, the rediscoverer of *Ramapithecus*, put it, the clock could not be correct, because *Ramapithecus* looked so much like a hominid, especially *Australopithecus*, in its jaws and teeth. How, in the first place, could anyone accept a molecularly derived date over a paleontologically derived date when the determination of that molecular date depended on securing at least one basepoint from the fossil record?

But Simons's comments carried a sobering message:

> If the immunological dates of divergence devised by Sarich are correct, then paleontologists have not yet found a single fossil related to the ancestry of any living primate and the whole host of species which they have found are all parallelistic imitations of modern higher primates. I find this impossible to believe ... [as] it is not presently acceptable to assume ... that *Australopithecus* sprang full-blown five million years ago, as Minerva did from Jupiter, from the head of a chimpanzee or a gorilla.

Sarich was well aware of Simons's comment. Sarich even excerpted this passage in a 1971 publication in which he summarized and defended his immunological methods. And in response to Simons's opening sentence — to the idea that paleontologists had "not yet found a single fossil related to the ancestry of any living primate" — Sarich appended a footnote. "This is silly, of course," he wrote. "*Australopithecus* is certainly ancestral to *Homo*."

But how do we know that *Australopithecus* is ancestral to *Homo*, or that *Australopithecus* has anything at all to do with *Homo*? By its molecules?

No. By its morphology.

If Sarich is going to use the fossil record to calibrate his molecular clock, and if he is going to accept the relatedness of *Australopithecus* and *Homo*, then he must admit the viability of morphology in reflecting phylogenetic relationships.

I suppose it was possible to accept the association of *Australopithecus* and *Homo* because it did not conflict with the five-to-eight-million-year period the molecular clock allowed for the divergence of the hominid lineage. It was thus perfectly acceptable for paleontologists to carry on about the morphological attributes that made the uniting of *Australopithecus* with *Homo* virtually unassailable. But *Ramapithecus* was another matter altogether. This apparent hominid — hominid for reasons as morphologically obvious as those that made *Australopithecus* a hominid — did not fit the molecular clock's timing of hominoid evolutionary events.

So, within the space of only a few paragraphs, Sarich turns one hundred eighty degrees and proclaims boldly, "To put it as bluntly as possible, I now feel that the body of molecular evidence on the *Homo-Pan* relationship is sufficiently extensive so that one no longer has the option of considering a fossil specimen older than about eight million years as a hominid *no matter what it looks like.*"

So much for morphology when it proves contradictory.

The sequence of events was as follows:

1. A phylogeny based on morphology was used to test the accuracy of immunologically and biochemically based theories of relatedness. There was general overall agreement in the arrangement of broader groups, with some disagreement on the placement of the occasional taxon. The criterion of overall similarity seemed appropriate in these endeavors.

2. Immunological and biomolecular techniques were applied broadly among organisms. Theories of relatedness were generated. Time scales were applied to these branching sequences.

3. When disagreement arose between a biomolecular or immunological phylogeny and one based on morphology, the latter theory was considered to be inaccurate in spite of the fact that the initial justification of the former molecular approach had come from an overall compatibility with a morphologically derived arrangement of taxa.

The result was an ever increasing acceptance of molecular and biochemical approaches to phylogenetic reconstruction. There were of course some remaining doubts and disputes, however, especially over two issues. One was whether molecular clocks, if they did exist, ran at regular rates. The other was whose technique yielded the more accurate phylogenies. There were disagreements not only between morphologists (especially paleontologists) and various molecular systematists, but between various molecular systematists themselves.

From the beginning, there was never total agreement in the arrangement of taxa in the phylogenies that were generated by different biochemical techniques. In their first paper, Sarich and Wilson disagreed with the results of other studies on the placement of the gibbon as well as the relationships of various lemurs and lorises. By 1976 the position of the gibbon seemed to be agreed upon. However, as is obvious from the papers in *Molecular Anthropology*, a 1976 volume edited by Morris Goodman and Richard Tashian, there still existed real differences between Sarich and Cronin (Sarich's more constant co-author during the 1970s) and Goodman and his co-authors in interpreting the relationships among the lemurs and lorises and in deciphering just where among primates the enigmatic tarsier belonged. There were other differences between the two biomolecular camps in how other prosimians were divided up.

Sarich and Cronin's albumin and transferrin data yielded a three-way split, from a common ancestor, of the mouse lemurs, the loris–bushbaby group, and the whole lemur–sifaka assemblage. Goodman and his co-workers nestled the mouse lemurs right in there with the lemurs and sifakas. Sarich and Cronin pictured a long period of separation between the bushbabies and the potto, whereas Goodman and his co-workers kept the bushbabies quite close to the potto.

These seemingly nitpicking discrepancies usually pass without much or any comment. After all, as a group, at least the typical lemurs, the sifakas, and the sportive lemur are united. This conforms to accepted dogma. The lorises and the bushbabies emerge as a group. This, too, conforms to accepted dogma. And the whole lot of them, along with the mouse lemurs, is clumped together as an even larger assemblage.

But these discrepancies do make a difference. Even if the picture of relationships was consistent with the generally accepted picture of the groups of prosimians — a picture which, I might add, is based on morphology — the discrepancies in lower-level relationships and detail still exist.

We might all agree that the typical lemurs, the sifakas, and the sportive lemur are somehow related more closely to one another than any of them is to a loris or a bushbaby, but are the typical lemurs more closely related to the sifakas or to the sportive lemur?

In a recent publication on the relationships among the lemurs and lorises, Ian Tattersall and I concluded that the best interpretation of morphology grouped the sifakas and the sportive lemur together. We also reiterated our longstanding position that the tiny nocturnal mouse lemurs, while confined now with the larger lemurs and sifakas to the island of Madagascar, are actually the evolutionary sisters of a loris–bushbaby group. This is more or less compatible with Sarich and Cronin's conclusions, which separate the mouse lemurs from the real lemurs, but at odds with Goodman's, which support the more bio-geographically based grouping of all primates now isolated on the island of Madagascar.

So what, you might think. It is, after all, just a matter of disagreement on the particular relations of only a few of the taxa under study. If there is general agreement on most of the higher-level associations, that should be sufficient.

Not so. These lower-level disagreements over who is most closely related to whom are the most relevant to matters of phylogenetic reconstruction. Perhaps because the taxa involved are prosimians, it does not matter as much just who really might be most closely related to the sportive lemur, unless you happen to be a sportive lemur. Try to shuffle around a genus or two within Hominoidea; the whole paleoanthropological world comes to a screeching halt.

The degree to which the relationships of a single taxon can be complicated is exemplified by the debate over the tarsier, that little primate from southeast Asia with big eyes, a cute face, and elongate tarsal bones. The tarsier has never enjoyed much systematic peace. Although *Tarsius* has commonly been thought of as being somehow intermediate between the "lower" primates (the lemurs and lorises) and the "higher" primates (the monkeys and hominoids), there have been decades of debate on just how close to one group or the other the tarsier really is.

The essence of Prosimii is that the tarsier lies closer to the lemurs and lorises. However, early in the twentieth century, the British anatomist R. I. Pocock pushed the tarsier closer to anthropoids and introduced a whole new set of problems into primate systematics. Phil Gingerich and I each went through a phase of arguing that *Tarsius*

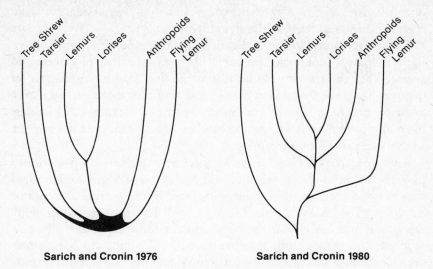

Sarich and Cronin 1976   **Sarich and Cronin 1980**

FIGURE 8D   Vince Sarich and Jack Cronin's refinement of their general scheme of primate relationships.

might somehow be related to the Paleocene plesiadapiforms, but recently I reviewed the systematic history of the tarsier and found no justification for any of these proposed arrangements. Instead, it seems that the affinities of *Tarsius* lie *within* Prosimii, specifically, with the loris–bushbaby–mouse lemur group.

Goodman attempted to unveil the phylogenetic relations of the tarsier through immunological analyses as well as the study of alpha and beta hemoglobin chains. In comparison with the slow loris, which was supposed to represent the prosimians, it appeared that the most straightforward association of the tarsier was as the sister of anthropoid primates.

Sarich and Cronin, in their 1976 contribution, arrived at a totally different arrangement of the primates as a result of their analyses of albumin and transferrin. The tarsier was excluded completely from the major primate groups. There was a lemur–loris group and an anthropoid group, both of which would seem to be justified on other criteria. However, *Tarsius* was not the sister of either of these assemblages. Rather, Sarich and Cronin came to the conclusion that the lemur–loris group and the anthropoid group had shared a common ancestor, which in turn was shared not just with the tarsier, but with the tree shrews as well as the flying lemurs.

Four years later, Cronin and Sarich added hemoglobin data to their immunological bailiwick and came up with a refined phylogeny. It seemed that the tree shrews had been the first of these mammals to diverge. Then the flying lemurs split off. And then came the ancestral primate. But here Cronin and Sarich remained steadfast in bucking the majority of morphologists and biochemists. The tarsier still emerged as the sister of all other primates.

Two years later, Goodman, with yet another set of co-workers, published a rejoinder to Cronin and Sarich. À la Sherlock Holmes, they entitled their rebuttal, which appeared in the journal *Systematic Zoology*, "The Case of Tarsier Hemoglobin." To set the record straight, Goodman and his colleagues went straight for the jugular: "In our opinion, the hemoglobin tree presented by Cronin and Sarich has been developed within a faulty conceptual and methodological framework and its repeated publication . . . perpetuates a serious misinterpretation of molecular data."

This accusation is not a minor one. When one is dealing with immunological reactions or electrophoretic banding patterns, there is some element of finesse in the interpretation of just how similar or dissimilar taxa really are overall. However, dealing with hemoglobins is another matter altogether.

Beginning in the 1960s, such molecules were being teased apart at a lower molecular level. The smaller molecules that made up larger ones, such as hemoglobin, were being identified, and the sequence in which these smaller molecules were laid down was being determined. And these sequences, whether they were of proteins such as hemoglobin or myoglobin, or of the fundamental molecules of inheritance, DNA and RNA, represented a kind of molecular morphology — a morphology whose units could be compared across taxa and used to identify individual taxa.

Macromolecules (large molecules), such as hemoglobin or DNA, are formed by smaller molecules that are aligned like beads on a string. In DNA and RNA, these smaller molecules are nucleotide bases, often simply called nucleotides. Nucleotides are arranged in units of three, called codons, with each codon specifying a particular amino acid. However, there is some redundancy in the code, so that more than one triplet of nucleotides can specify the same amino acid. A

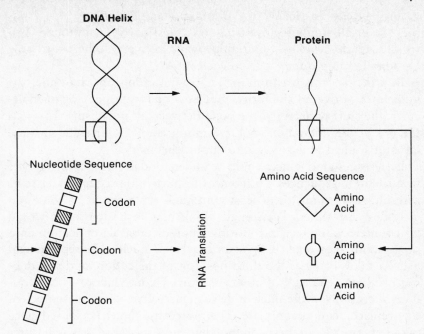

FIGURE 8E   A simplified representation of protein synthesis: information from DNA (the sequence of nucleotides in the DNA) is encoded in a simple strand of RNA, which serves as the template for organizing a particular sequence of amino acids, which form a protein.

string of amino acids, coded for by a sequence of codons, makes up a protein, such as hemoglobin, albumin, or any of the blood serum proteins.

The molecular sequence one would like to identify is the sequence of nucleotides, because this is at the level of DNA. Sometimes, however, one can get only to the level of amino acid sequences, which, because of the redundancy in the code, must be viewed as an approximate reflection of the sequence of nucleotides. Even with this potential error in going from protein sequences to an extrapolation of nucleotide sequences, it's reasonable to pursue such data. After all, you can't reconstruct a nucleotide sequence from the study of teeth and bones.

In 1962, Emile Zuckerkandl and Linus Pauling formulated one of the first evolutionary explanations of molecular change, using hemoglobin as the example. What Zuckerkandl and Pauling proposed was that "over-all similarity must be an expression of evolutionary history," with descendants "mutating away" from each other, becoming "gradually more different from each other."

In principle, this was nothing new. Who didn't operate under the guidelines of overall similarity? But now one was speaking of molecules, the stuff of heredity, the bond between all organisms. This was the era of the "new biology." Evolution could be brought into the lab and studied objectively through biochemistry.

Differences among taxa could be understood in terms of differences in nucleotide sequences. Differences in nucleotide sequences arise as a result of constant, or at least continual, replacement of some nucleotides with others. You might think that such alterations would be counterproductive, introducing unpredictable complication into the system. However, it appears that such nucleotide replacement usually takes place at the third position of the codon, and this does not produce "change" of any evolutionary consequence — only minute alterations in the nucleotide sequence. Over time, these nucleotide replacements accumulate. The more time that has elapsed since taxa separated from their common ancestor, the more they will differ at the molecular level — they will have fewer stretches of their nucleotide sequences in common.

So, as a species is continuing along through time, generation after generation, the sum of its molecular baggage is changing as a result of nucleotide substitution. All species are continually becoming increasingly different from their sister species. If a species were to become the ancestor of two new species, these two new species would, over time, accumulate nucleotide replacements as they, at the genomic level, continued to diverge from their ancestor's nucleotide sequences.

The idea that "over-all similarity" is a direct reflection of "evolutionary distance" derives from the premise of accumulated nucleotide change in descendants. However, this particular notion exists in part only because there appeared to be a correlation between (hemoglobin) differences and an evolutionary arrangement of taxa as traditionally viewed. It was a necessary assumption. As Zuckerkandl and Pauling wrote, their observations "can be understood at once if it is assumed" — I would emphasize "assumed" — "that in the course of time the hemoglobin-chain genes duplicate, [and] that the descendants of the duplicate genes 'mutate away' from each other."

On the face of it, this makes some intuitive sense. But there is still no reason to exclude the possibility that, having separated from a common ancestor, the descendants would carry on at different molecular rates of change. This, of course, would not yield as nice a picture as if everything proceeded in the same way among all taxa.

1 Real evolutionary relationships of species

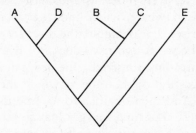

2 Real evolutionary relationships of species illustrating overall degrees of similarity

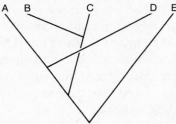

3 "False" scheme of relationships of species based on overall degrees of similarity

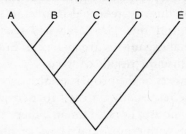

FIGURE 8F Although the real evolutionary relationships of various species may be as in diagram 1, an adherence to using overall similarity as the criterion for determining closeness of relatedness can yield an incorrect scheme of relationships (diagram 3), especially if one or more of the species (for example, D) is wildly different from the others because of its own peculiarities (diagram 2).

If molecular change occurred differentially, a species might even be more similar to an earlier branching relative than to its own evolutionary sister.

Where did the notion that "over-all similarity" was "an expression of evolutionary history" come from? Zuckerkandl and Pauling stated that this conclusion was "indicated by the gradually increased amount of differences found when human hemoglobin is compared with hemoglobins from progressively more distant species." There is nothing at all in their work, however, which indicates to me the rate at which any difference arises. For example, the discovery of a hundred nucleotide differences between two taxa does not translate into the rate at which the substitutions occurred. One needs additional information, such as a suggestion of the time at which these differences started to accumulate. And then one has to assume that the rate of substitution over this period of time was gradual and that differences did not arise as a result of a few concentrated phases of replacement.

Zuckerkandl and Pauling compared the hemoglobin sequences of humans and fish, on one end of the evolutionary scale, and humans and gorillas at the other extreme. A comparison with the horse was also added. At this level of comparison, there did seem to be a rough agreement between hemoglobin sequence differences among these taxa — the fish being more dissimilar to humans than the horse, and the gorilla the most similar — and the accepted evolutionary arrangement and distancing of these animals.

This intriguing but meager "discovery" was, however, elevated to the status of a general "law" of "genetic behavior." Relatedness was supposed to be directly reflected in overall molecular and biochemical similarity. And this "law" always seems to be corroborated when additional organisms are analyzed. But this "corroboration" exists only because the results of whatever technique is used are interpreted in the context of the properties already established for the genome, and because, especially on the broader levels of animal evolution (fish, chicken, mouse, horse, human), there does appear to be reasonable agreement with traditional views of phylogenetic arrangements.

If this "law" is real, then there should not be the amount of disagreement that exists between various molecular and biochemical studies on the evolutionary position of certain taxa. The tarsier is a case in point. However, when such discrepancies are uncovered, they

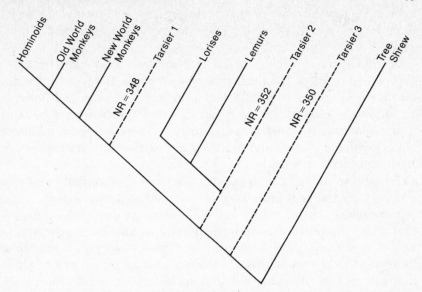

FIGURE 8G   Three alternative positions of the tarsier among primates derived
from analysis of hemoglobin nucleotide substitutions (NR). The three schemes
are essentially equally viable, the choice among them being decided on the basis
of other, nonbiomolecular information.

are not viewed as symptomatic of a problem with the general underlying assumptions — the "law." Rather, these discrepancies are thought
to exist for another reason, perhaps as the result of "a faulty conceptual and methodological framework" on the part of the contrary
investigator or investigators.

What is the case of the tarsier's hemoglobin? You would think that
there really was only one possible answer from the data — and that
this was the correct answer. However, the truth of the matter is
that Goodman and his co-workers, in trying to construct the most
parsimonious phylogenetic scheme (the most parsimonious being determined by the fewest postulated nucleotide changes) actually arrived
at three different arrangements, which differed from each other by
only a few nucleotide replacements.

Each hemoglobin chain — alpha and beta chains in the adult and
gamma in the fetus — is composed of almost 150 amino acids. Since
three nucleotides make up the codon that codes for each amino acid,
a hemoglobin chain corresponds roughly to just under 450 nucleo-

tides. The goal is to arrive at a phylogeny that has its taxa arranged in such a fashion that the nucleotide difference between them is minimal.

A phylogeny in which *Tarsius* is positioned as the sister of Anthropoidea requires 348 nucleotide replacements. If *Tarsius* is shifted so that it is the sister of all other primates — Cronin and Sarich's hypothesis — two additional nucleotide replacements, or a total of 350 replacements, must be stipulated. If *Tarsius* is grouped most closely with the lemur and slow loris sampled, the number of necessary nucleotide replacements increases to 352.

Strictly speaking, the tree with the fewest nucleotide replacements — the one that links *Tarsius* with Anthropoidea — is the most parsimonious. But in reality the three trees are essentially indistinguishable: the differences in nucleotide replacements are not statistically significant. And Goodman and his co-workers admit this. "However," they go on to say, "the real meaning and value of alternative trees with different NR [nucleotide replacement] scores does not emerge from a simple statistical test, but from an enlightened analysis of complex and dynamic data sets whose changing parameters constantly influence the meaning derived from those scores."

What does an "enlightened analysis" consist of?

An "enlightened analysis" includes "data derived from . . . the fossil record" as well as other sources.

Why? — especially if molecular similarity is so revealing?

The reason, Goodman and his colleagues go on to state, is that "molecular anthropologists are generally skeptical, for example, of lowest NR trees that save a few NR's but depict relationships among lineages for which there is very little or no *a priori* support from comparative anatomy or the fossil record."

But now we are back to square one. The "best" molecular phylogeny is the one that fits best with the morphological data. How is this an advance in method, precision, and objectivity?

Furthermore, in those instances when a molecular phylogeny differs from a morphological phylogeny, how can one presume that the molecular phylogeny is the correct one, or even the preferable one?

I don't see how one can.

This issue is relevant to the latest study on hemoglobin sequences among the large hominoids. Here, Goodman and yet other colleagues demonstrated that, in keeping with other data that show humans and the chimpanzee to be almost indistinguishable at the molecular level, the hemoglobins of humans and chimpanzees were also identical. The two adult hemoglobin chains — alpha and beta — were analyzed. Goodman and his colleagues found that the common chimpanzee and its sister species, the pygmy chimpanzee, had identical hemoglobin amino acid sequences. And this sequence was identical to a human's. In comparison with these hominoids, the gorilla differed in its hemoglobin sequence by two amino acid substitutions, and the orang-utan by only three more.

In terms of the reconstruction of molecular events, the orang-utan was considered to have changed at one position in its hemoglobin subsequent to its divergence from the last ancestor of the large hominoids. Two changes occurred in the lineage leading to the African apes and *Homo*. The gorilla acquired an additional amino acid substitution after it diverged, while the line leading to the human–chimpanzee group also acquired a substitution. No subsequent alterations of hemoglobin sequences occurred after the splitting of the human line from that leading to the two chimpanzees, or after the divergence of the two species of chimpanzee.

If molecules are supposed to change continually, why, you might wonder, is there no difference between humans and the two species of chimpanzee?

The explanation offered for why there was no change following the divergence of the gorilla is that hemoglobin, in contrast to some other macromolecules, is supposed to evolve slowly. There was not enough time following the divergence of humans and then the two chimpanzees for alterations in amino acid sequences to be brought about. Thus, study of hemoglobin sequences demonstrates not only the close relationship of humans and the chimpanzee, but the youth of the human lineage. The split between the human and chimpanzee lineages must be quite recent, and not of the time depth required to embrace *Ramapithecus* as a human relative.

In a much more comprehensive article, Goodman and his co-workers took into consideration the available data on alpha and beta hemoglobins, myoglobin, fibrinopeptides, and a few other proteins. The gibbon was included in this study. From this enlarged data base, Goodman and colleagues concluded that more than one parsimonious phylogeny could be constructed. "It is equally parsimonious with these amino acid data," they wrote, "to switch the positions of *Hylobates* [the gibbon] and *Pongo* or, as yet another alternative, to join *Pongo* to *Hylobates* and then join this branch to that of *Pan, Homo* and *Gorilla.*" The phylogenetic arrangement that was illustrated, however, was the one where humans and the chimpanzee are depicted as sister taxa.

I wonder what the results would have been had the orang-utan been positioned nearest to *Homo* and the consequent parsimony, or lack of it, calculated. Since the number of amino acid differences in hemoglobin sequences in general among the hominoids is about as restricted as in *Tarsius* when this primate is associated with different groups of primates, I am reminded of the argument of applying an enlightened analysis to the resolution of statistically indistinguishable phylogenies.

In addition to hemoglobin, it has been popular to use myoglobin — a blood serum protein in the same oxygen-carrying family (globins) as hemoglobin — in support of a human–African ape, if not just human–chimpanzee, alliance. The group of researchers that did the myoglobin analyses, encompassing molecular biologists, biological mathematicians and statisticians, and an evolutionary morphologist and paleontologist, was based in Cambridge, England, and headed by A. E. Romero-Herrera. As Ken Joysey, the paleontologist of the group, told me once, they felt that a collection of collaborators of different disciplines would provide a series of healthy checks and balances throughout the whole process of sequencing the myoglobin and, especially, of interpreting the results.

Their first paper, published in *Nature* in 1973, was a presentation of the initial myoglobin sequences obtained at that time and an application of these sequences to the matter of deducing the phylogenetic relationships of the organisms involved. The animals for which Romero-Herrera et al. had worked out myoglobin sequences included a marine worm, lamprey (the eellike parasite that attaches itself to sharks), kangaroo (representing the marsupials), ox and sheep (representing the even-toed grazers), horse (an odd-toed grazer), seal (a carnivore), whale, porpoise, dolphin, and various primates. The number of primates sampled equaled the total of all the other mammals.

The general "evolutionary arrangement" of these organisms was in keeping with the basic groupings otherwise accepted for these animals. For example, the ox and sheep were united and then, as a group, they were linked to the horse. The whale, dolphin, and porpoise fell out as a cluster. And all the primates, as a group, sorted out together.

Among the primates, the sportive lemur and the bushbaby, representing the prosimians, clustered and were separated from the anthropoid taxa. The marmoset, woolly monkey, and squirrel monkey clustered as the New World monkey group. The baboon and macaque came together as the Old World monkeys. And the gibbon emerged as the sister of the human–chimpanzee duo. Thus, albeit with minimal representation of taxa, prosimians and anthropoids emerged as expected, and within the hominoid group, the chimpanzee emerged as most similar to *Homo*.

In 1976, again in *Nature*, Romero-Herrera and his co-workers published the myoglobin sequence for the orang-utan. Their results were so unexpected that they entitled their article "Myoglobin of the Orangutan as a Phylogenetic Enigma." The upshot of their analysis was that the most economical and parsimonious interpretation of the myoglobin data yielded an arrangement among the hominoids in which the gibbon came out as closer to humans and the African apes than did the orang-utan. Only by allowing for a more complicated scheme of myoglobin "evolution" among the hominoids — with, for instance, various "back-mutations" — could one arrive at the more commonly accepted arrangement: gibbon first, then the orang-utan, and then the human–African ape group. In their published pondering of this "enigma," Romero-Herrera and his colleagues cited an earlier study by the Goodman group, in which it was reported that "with rabbit anti-human albumin serum, gorilla and chimpanzee seem to be almost identical to man while gibbon and siamang show slight divergence and orangutan shows yet more divergence." There must have been some solace in the knowledge that they were not the only molecular systematists to come up with so untraditional a phylogeny.

Although many phylogenetic schemes among the hominoids were considered by Romero-Herrera and his co-workers, they do not appear to have entertained the grouping of humans with the orang-utan, despite their acknowledgment that "the amino acid sequence of the myoglobin of the orangutan differs from that of man at only two positions." At position 110 in the myoglobin sequence, the orang-utan has the amino acid serine, and in this regard resembles the Old World monkeys. Gibbons, the African apes, and *Homo* have the amino acid cysteine at position 110, and in terms of amino acid

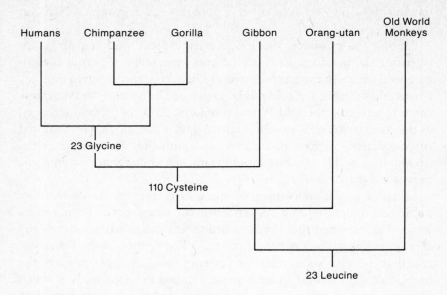

FIGURE 8H   The "enigma" of the orang-utan's blood serum protein myoglobin: the seemingly most parsimonious interpretation of the data is that the gibbon is more closely related to humans and the African apes than is the orangutan, which is contrary to the popular biomolecular as well as morphologically based schemes of relationships.

transmutation, cysteine appears to be derivative of serine. Because the orang-utan resembles the Old World monkeys in having serine at position 110, a logical conclusion (and the one that generated the "enigma") is that the four hominoids with cysteine are united as a clade to the exclusion of the orang-utan. Any other arrangement of the hominoids requires the "re-evolution" of serine at position 110 in the orang-utan.

In the prosimian primates and all other mammals, the amino acid at position 23 of the myoglobin sequence is glycine. Among anthropoids, humans and the African apes also have glycine at position 23. The orang-utan, the gibbon, and the monkeys have serine at position 23. Romero-Herrera et al. concluded that a hypothesized common ancestor of humans and the African apes could have "re-evolved" glycine at position 23, or that humans and the African apes could have independently "re-evolved" glycine at position 23. But the most straightforward conclusion — following the logic that placed the gibbon closer than the orang-utan to the other hominoids — was that having glycine at position 23 is primitive for mammals and anything else at position 23 is the evolutionary novelty. Thus, orang-utans, gibbons, and monkeys would be "united" by their possession at position 23 of serine. In any case, and regardless of how one interprets the distribution among the hominoids of the different amino acids found at positions 23 and 110, the myoglobin sequence data do not yield the typical and comfortable arrangement of humans and the African apes.

The most frequently cited paper by Romero-Herrera and his crew on myoglobin is their publication in the 1978 series of the *Philosophical Transactions of the Royal Society of London*. In this book-length article, Romero-Herrera and his colleagues presented a thorough discussion of the myoglobin molecule, their methods, and an augmented sampling of animals. There were now chicken and penguin, opossum was added to the marsupial group, dogs and badger beefed up the carnivores, zebra joined the horse, and potto and slow loris expanded the ranks of the prosimians. The tree shrew and hedgehog represented newly added groups.

From myoglobin data, everything seemed to agree with the general arrangement of these taxa and their groups, including the hominoids. In one scheme, the gibbon was depicted as the sister of a group which had humans united with the African ape duo. In another calculation, the African apes were broken up, with the chimpanzee falling out as the sister of *Homo*. But the orang-utan was not included in this publication. One wonders what the addition of this myoglobin data would have done to the other calculations of relatedness.

Although the orang-utan has quite often not been included in many of the published molecular and biochemical studies, often for lack of data, there has nevertheless been a growing belief that these studies really have demonstrated the extreme evolutionary closeness, to the exclusion of any other hominoid, of humans and one or both of the African apes.

I am, however, uncomfortable with analyses that do not include at least all potential members of the group under investigation, and would prefer to see as many different animals as possible included in the comparison. How else can one attempt to rule out parallelism as a possible reason for the similarity one sees among organisms?

But a common sentiment in molecular and biochemical work is that, especially because so many molecular bits are being sampled, the possibility that much similarity is due to convergence or parallelism is minimal. And because molecular change — nucleotide replacement — is supposed to be essentially unidirectional and additive, similarity between taxa will reflect their genomic change away from the more primitive state. It is for this reason that what I consider to be incomplete samples are nevertheless believed to yield phylogenetically revealing schemes of relatedness.

On the face of it, there appears to be some logic to the argument. If humans and chimpanzees emerge from any given analysis as virtually identical, then every other taxon (even if it has not yet been sampled) must be more different. However, this does not then necessarily mean that an evolutionary relationship has been established. The work has only just begun.

9

Of Molecules
and Evolution

SIMILARITY AND GENEALOGY — those are the keys to molecular
approaches to evolution. Similarity is genealogy and genealogy is
similarity. This equation is so ingrained in many systematists' minds
that one need only address similarity for genealogy to be assumed.

This was the case, for instance, with a review article by Mary-
Claire King and Allan C. Wilson (of Sarich and Wilson) that appeared
in *Science* in 1975. King and Wilson presented all available data on
blood serum proteins, as well as the results of DNA hybridization (a
technique I will soon discuss). The upshot was that humans and
chimpanzees differed in genetic makeup by only about one percent.
King and Wilson concluded that "all the biochemical methods agree
in showing that the genetic distance between humans and the chimpanzee is probably too small to account for their substantial organismal differences." In order to explain how humans and chimpanzees
could be virtually identical in their genes but markedly different animals, King and Wilson suggested that humans and chimpanzees must
be different in those genes which regulate development.

The two major types of genes are structural and regulatory. Structural genes code, for instance, for proteins, elements that together
make up the bits and pieces of an organism. Regulatory genes control
the expression and interaction of structural genes. They regulate the
timing of development.

A simple analogy can be found in the different rates at which boys
and girls develop, girls reaching physical maturity much earlier than
boys. Boys and girls follow the same developmental path until shifts

in hormonal secretion cause boys to veer off on a course toward masculine features, girls toward feminine features. Of course, males and females of the same species are nowhere near as different from each other as members of different species are. However, the regulation of development with a re-orchestration of the interactions of the basic building blocks can apparently lead to vastly different organismal end products.

King and Wilson felt that differences in regulatory genes could account for the visible differences between humans and chimpanzees. Although King and Wilson stated that "the only two species which have been compared by all of these methods are chimpanzees . . . and humans," and thus "a good opportunity is . . . presented for finding out whether the molecular and organismal estimates of distance agree," their study is typically cited as demonstrating the extreme evolutionary closeness of humans and chimpanzees to the exclusion of other hominoids.

Four years later, in 1979, Elizabeth Bruce and Francisco Ayala, of the University of California at Davis, published in the journal *Evolution* their analyses of blood serum proteins. In their discussion, Bruce and Ayala laid out the ground rules: "Information macromolecules — i.e. nucleic acids and proteins — document evolutionary history . . . [thus] degrees of similarity in such macromolecules reflect, on the whole, degrees of phylogenetic propinquity." The more similar the molecules, the closer the relationship. But they go on to say, in an apparent paraphrase of Goodman's enlightened approach to systematic issues, "The information gained from the study of macromolecules must be weighed together with that derived from comparative anatomy and paleontology in order to ascertain phylogenetic relationships among organisms."

From the electrophoretic data alone, Bruce and Ayala concluded that "man, the chimpanzee, the gorilla, and the orangutan are about equally related to each other." Bruce and Ayala did not discriminate further the possible relationships among the large hominoids. However, the phylogenetic scheme they favored, based on their evaluation of the "molecular as well as morphological and paleontological" evidence, depicts humans and the African apes united as a group to which the orang-utan and then the gibbon are related.

In the course of rereading Bruce and Ayala's article for a publication I was preparing, I decided to examine their tables of calculated genetic

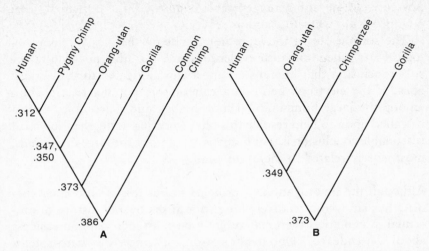

FIGURE 9A The different schemes of relationship among the living large hominoids, using the raw data presented by Elizabeth Bruce and Francisco Ayala. In diagram A, the pygmy chimpanzee and its apparent sister species, the common chimpanzee, are dissociated. The relationships in this diagram and in diagram B certainly contradict the commonly accepted schemes derived from other biomolecular as well as morphological studies.

distances and see what phylogenies would be generated if these data were considered alone. From one of their tables of genetic distances, calculated according to one particular method, the following sequence emerged: the pygmy chimpanzee was the closest to *Homo*, the next closest was the orang-utan, followed by the gorilla, and then the common chimpanzee. This was unexpected, because not only does this "phylogeny" not have any resemblance to anything ever suggested from any source, the two apparent species of chimpanzee do not even cluster together. The genetic distances Bruce and Ayala presented in another table, and which they calculated using a different method, placed the orang-utan and the chimpanzee equidistant from *Homo*, with the gorilla falling out as the sister of these three hominoids.

It may be that the genetic distances between these hominoids, regardless of the method of calculation, are statistically indistinguishable. If this is the case, these genetic distances do not lend themselves to the support of any one scheme of relatedness among the large hominoids over any other. Taken at face value, these distances do not corroborate any arrangement among the large hominoids that anyone would want to claim as reality.

The sorting out of the large hominoids on the basis of morphological and paleontological evidence would certainly not lead to the sole conclusion that humans are most closely related to the African apes. If the electrophoretic data cannot sort out the relationships among the large hominoids, if morphology and paleontology must be called upon to help resolve this phylogenetic puzzle, then the most reasonable conclusion is that humans are, among the large hominoids, most closely related to the orang-utan.

Although the investigation of proteins to buttress evolutionary theories has enjoyed a relatively long stint in the overall history of molecular systematics, there has recently been an increasing uneasiness about the endeavor. One major reason for concern is that proteins are a step away from the real "stuff," the DNA. It is DNA which codes for the amino acid sequence that makes up a given protein. And since we know that there is some redundancy in the genetic code — that more than one arrangement of a codon or triplet of nucleotides can code for the same amino acid — what appear to be similar proteins (which are just amino acid sequences) may have been generated by slightly different DNA sequences.

The identification of DNA nucleotide sequences can be accomplished much better now than even a decade ago, but it is still a long and laborious task. There is, however, another way in which to get information at the level of DNA. It's not as direct and doesn't reveal the identity of the DNA molecule. But it's much easier than sequencing.

Within DNA there are only two large nucleotides and two small nucleotides. A single strand of DNA is a string of these nucleotides encoding certain proteins.

A DNA molecule is composed of two strings of nucleotides. Each strand is a molecular complement of the other, with a large nucleotide on one strand matched by a smaller nucleotide on the other. A chemical bridge connects each complementary pair, like a rung of a ladder. However, rather than looking like a ladder, with straight sides, the double-stranded DNA molecule is twisted into the shape of a helix.

When it is necessary to produce a protein, the DNA helix "unzips" along the appropriate section. Molecules available within the cell's center, or nucleus, are attracted to the open section of DNA and arrange themselves into a chain complementary to one of the strands of unzipped DNA. The product is a strand of RNA, in this case "mRNA," or messenger RNA, because it will carry the coded sequence for the necessary protein to another part of the cell, where different RNAs will put together the amino acids in the correct sequence.

It does not make any difference which strand of the unzipped DNA the mRNA "copies." The large nucleotides will always link up with the small nucleotides. And only one type of large nucleotide will link up with one type of small nucleotide. So the sequence on one strand of DNA has to generate the sequence of the other strand of the DNA molecule.

Let's say that you took one of your DNA molecules and unzipped its two strands down the middle, but then you changed your mind — you now wanted to put the DNA molecule back together again. You put the two dissociated strands in a test tube, hoping that these two strands would reassociate in the correct alignment and with complete complementarity. And they would. There would be no gaps between the two strands at places where they did not match up because these two strands came not only from the same species of animal but from the same individual. The complementary nucleotide sequences would essentially be homologous.

If you were to take one of your DNA strands and a strand of DNA from another organism, you would probably predict there would be gaps where the sequence of nucleotides on the two strands would not match up. There might be sections where a large nucleotide might oppose a large nucleotide and a small nucleotide might oppose a small nucleotide. To explain this you might suggest that the degree to which the strands differed would be a reflection of the degree to which the organisms under study were distanced evolutionarily from each other. The greater the similarity in nucleotide sequence, the more closely related the organisms.

But exactly to what degree are the DNAs of animals similar or different? It would be ideal to identify the actual DNA sequences and the specific elements of the sequence which differed between organisms, but this is exceedingly tortuous work. One can get an overall idea of complementarity, however, by measuring the heat it takes to dissociate the two strands of DNA. A major problem with this shortcut is that you can't tell where and in what order the differences between the strands occur, or the nature of the difference.

If you heated one of your DNA molecules, it would take a high

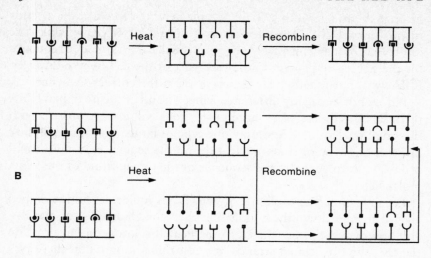

FIGURE 9B A simplified illustration of DNA hybridization. Heat, in all cases, is used to decouple the two strands of a DNA molecule. In diagram A, the decoupled strands of an individual's DNA will reassociate perfectly. However, when decoupled strands of DNA from two different individuals are placed together (diagram B), they will be capable of forming a hybrid double-stranded DNA molecule only along those segments of each strand that bear compatible nucleotides.

temperature to break the bonds that hold the two strands together. If you were to "make" a double-stranded DNA molecule from one of your DNA's strands and a strand of DNA from another animal, it would take less heat to break apart the two strands. Since no other animal's DNA sequence is identical in every respect to your own, there are some gaps in the molecule, places where no "ladder rungs" hold the strands together. The less complementarity between DNA strands, the less strongly held together the molecule is, and the less heat it takes to disrupt its bonds.

The technique of cleaving DNA double helices and combining DNA strands from different organisms is called DNA-DNA hybridization, or just DNA hybridization. The degree to which a DNA hybrid is thermally stable reflects the complementarity of the strands.

One of the first applications of this technique appeared in a study by R. J. Britten and D. E. Kohne that was published in *Science* in 1968. Their main concern was to try to find out more about the genome. What they discovered was that the genome of higher organisms — but not of bacteria or viruses — contained "hundreds of thousands of copies of DNA sequences." These repeated sequences made up a considerable portion of the genome. What was more surprising than their profusion was that these repeated DNA sequences were "trivial and permanently inert." Only a small fraction of the genome was composed of unrepeated, unique, single-copy DNA sequences, which apparently constitute the "guts" (or, perhaps more properly, "brains") of the genome.

In discussing genomic change, Britten and Kohne entertained the notion of two very different mechanisms. One was that changes — here in nucleotide sequence — occur slowly over time. This model would explain only the "divergence of pre-existing families" of nucleotide sequences, not the introduction of new families of sequences. In the second mechanism, new sequences "result from relatively sudden events," which are called "saltatory replications." According to Britten and Kohne, "saltatory replications of genes or gene fragments occurring at infrequent intervals during geologic history might have profound and perhaps delayed results on the course of evolution." This then is a general mechanism by which genomic differences can be generated.

How many genomes studied to date, and thought to contain some number of single-copy DNA sequences, contain single-copy DNA that will eventually become repeated DNA later on, after that particular species has been in existence for a while? We can only recognize the existence of a family of repeated DNA sequences in retrospect, after multiple copies of the founding genetic generation have arisen. It may be that families of repeated DNA are only residuals of changes that were initially significant at the origin of new taxa. Divergence need not be inexorably slow. Difference could arise through a more "catastrophic" route — a genomic "punctuation" or "pulse" — which catapults its bearer species temporarily off the conveyor belt of slow, gradual change. If we always assume that gradualism is the only mechanism for achieving change, we will never recognize the genomic oddball for what it is. Its "difference" will always be interpreted in the usual way, as a reflection of evolutionary distance, where difference equates with the negative connotation of "farther away phylogenetically." In reality, the oddball organism itself would have become different from the others phylogenetically around it, leaving its nearest kin similar, but not any more closely related, in their world of un-eventful genomic change.

One of the early attempts at using DNA hybridization to recon-struct relationships among the primates appeared in *Nature* in 1976. Raoul Benveniste and George Todaro's main focus was on the dis-tribution among mammals of the type C viral gene. Benveniste and Todaro discovered that the presence of the type C viral gene was particularly characteristic of animals found in Asia, including the orang-utan and the gibbon. Type C viral genes were not found, for example, in African mammals, such as the chimpanzee and the gorilla. To Benveniste and Todaro's surprise, humans turned out to have type C viral genes, which suggested that humans had had a long period of "evolution," or at least residence, in Asia.

Given the focus on Africa as the seat of human evolution, the rekindling of the suggestion that human evolution had occurred in Asia was not taken well. I have seen the Benveniste and Todaro paper cited as claiming that the presence of type C viral genes in humans and the orang-utan meant that the two were closely related, and the publication ridiculed for this claim. Benveniste and Todaro made no such assertion. The C viral gene data themselves do not warrant it. Benveniste and Todaro did attempt to use DNA hybridization data to reconstruct the relationships of the catarrhine primates, however. In terms of DNA hybridization, the Old World monkeys fell out as a group, as did the hominoids. Among the hominoids, the thermal instability of the "manufactured" DNA hybrids indicated that the gibbon and siamang were the most different of all. Among the large hominoids, the percentage of sequence mismatch was lowest when *Homo*, chimpanzee, and gorilla were compared. The orang-utan's DNA formed a somewhat less stable hybrid with *Homo*.

To calculate their evolutionary "numbers," Benveniste and Todaro relied on the conclusions of previous studies, since it is necessary to know how differences in mismatches between DNA hybrids translate into differences in thermal stability. You need a formula that will tell you how great or small the nucleotide mismatch is if you supply the information on how much heat it took to break up the DNA hybrid.

According to Benveniste and Todaro, "the effect of mismatched base-pairs on thermal stability" is "between 0.7 and 1.7°C per 1% altered pairs." Since nuclear DNA consists of anywhere from 10 to 100 million nucleotide pairs, every one percent of difference in nucleotide sequence between DNA hybrids will result in a drop of about one degree centigrade of the heat necessary to melt or cleave the hybrid molecule.

First, one determines the temperature needed to melt a natural, untampered with, and unhybridized DNA molecule. Let's take, for instance, a DNA molecule of a baboon. A certain amount of heat is required to break down a baboon's DNA molecule into its two strands. Then one makes a hybrid molecule from a strand of baboon DNA and a strand of another animal's DNA — for example, the DNA of a chimpanzee. By measuring the temperature at which the DNA hybrid melts, one can calculate the difference in melting temperature between the "real" DNA from the baboon and the hybrid DNA. If a 0.7–1.7°C difference in temperature corresponds to approximately a one percent difference in the sequence of nucleotides, then, with relative ease, one can calculate the approximate percent of sequence mismatch between the standard animal and those with which its DNA is hybridized.

When hybrids were made of various combinations of human, chimpanzee, and gorilla DNA, Benveniste and Todaro found that melting temperature differences ranged between 2.3 and 2.5°C. By multiplying the low end of the range (2.3°C) by the lowest percent mismatch (0.7%) found among these three hominoids, and by multiplying the high end of the range (2.6°C) by the highest percent mismatch (1.7%), one derives a range of 1.61% to 3.42% mismatch in nucleotides among humans, chimpanzees, and gorillas. DNA hybrids with the orang-utan yield a difference in melting temperature of 4.5°C, or a range of 3.15% to 7.65% nucleotide mismatch. The gibbon and siamang melting temperatures were quite a bit higher, with a 6.3–6.5°C difference.

This study is often cited as demonstrating the unity of humans and the African apes and separating these three hominoids from the orang-utan. This reading — at least the one about separation — does seem warranted on the basis of the differences in melting temperatures. Andrews and Cronin, in their review in *Nature* on the evolution of the orang-utan, summarized Benveniste and Todaro's data as indicating a 1.1% nucleotide mismatch between *Homo*, the chimpanzee, and the gorilla, and a 2.4% nucleotide mismatch between *Homo* and the orang-utan. However, the calculations I just went through — considering on face value the allowable temperature ranges as well as the allowable margin of error in percent of nucleotide mismatch per degree centigrade of temperature difference — indicate that there is a bit of overlap among all of the hominoids.

If, in summary, humans and the African apes do indeed differ by only about 1.1% in their DNA, and humans and the orang-utan differ by only about 2.4% in theirs, is the difference between 1.1% and 2.4% statistically meaningful? How significant can this difference really be when you take into consideration the fact that a DNA molecule may consist of 10 to 100 million nucleotides? But even if we accept for the moment that these differences are statistically meaningful, we still do not know what the difference means. After all, one cannot locate and identify those aspects of the sequence which are different and thus one doesn't know if difference means the animal in question is more primitive or actually more specialized than those to which it is being compared.

Nevertheless, one of the latest crazes in molecular systematics is the use of DNA hybridization in resolving the evolutionary relationships of organisms. Some evolutionary biologists believe that finally there is a technique which is problem- and error-free and which generates the "real" phylogenies.

In reviewing the earlier studies on DNA hybridization, Charles Sibley and Jon Ahlquist, two current champions of the technique, concluded that a major problem in all of them was that only a limited number of cross hybridizations had been tested. This, claimed Sibley and Ahlquist, was why those studies had so much difficulty in refining the relationships among humans and the African apes — usually these three hominoids were grouped, but the level of resolution was never fine enough to determine which two were the most similar (and thus the most closely related) of the three.

Sibley and Ahlquist felt another source of error in these earlier studies came from forming hybrids with all of an animal's DNA, repeated as well as single-copy. They argued that it was important to remove all repeated DNA — since it was not only redundant but not the real "stuff" — and to work only with single-copy DNA.

Before they tackled the relationships of hominoid primates, Sibley and Ahlquist spent years applying their DNA hybridization technique to the phylogeny of birds. Their efforts were ambitious and eventually produced phylogenies for all the major groups of birds. They also developed the notion of a uniform average rate of genomic change

(UAR), which they felt characterized the nature of evolutionary change across the taxonomic board.

Sibley and Ahlquist argued that the power of DNA hybridization in "discovering" the phylogenetic relationships of organisms comes from the entire genome of an animal being sampled. They claimed that the "law of large numbers" (the entire genome is composed of millions upon millions of nucleotides) provides the checks and balances necessary to rule out "false" similarities or parallelisms. Indeed, since DNA hybrid strands will only link up along those stretches of sequences which are complementary, there is no question about homology. If the sequences do not line up, those portions of the strands are not homologous. And that is that.

The notion of there being a uniform average rate of change also takes care of debates over constant versus irregularly running molecular clocks. As Sibley and Ahlquist have argued, there appears to be a uniform rate at which the entire genome changes, even though the rates for different molecules may differ.

Sibley and Ahlquist's melting temperatures for the DNA hybrids did produce numbers that could be converted into phylogenetic distances. To make the conversion, they used a procedure called "average linkage." "Average linkage" begins "by clustering the closest pair or pairs of taxa," after which "one links the taxa which have the smallest average distance to any existing cluster," and on it goes until "all taxa are linked." The underlying assumption, which permits the linking of the most similar pairs, is that the DNA hybridization technique "measures the net divergence between the homologous nucleotide sequences of the species being compared."

Many aspects of Sibley and Ahlquist's phylogenies on birds were consistent with phylogenies generated through morphology. A biologist at the University of Pittsburgh, Robert Raikow, has used the anatomy of the hindlimb musculature as a source of data for reconstructing the relationships among various bird groups and feels that his and Sibley and Ahlquist's phylogenies are quite compatible. But there is not always total agreement between morphological and molecular phylogenetic schemes. Mary McKitrick, a recent doctoral student of Raikow's, disagrees with Sibley and Ahlquist on the broader phylogenetic relationships of, and the details of relatedness among, the flycatchers.

I first met Charles Sibley at a special symposium for an annual meeting of the American Association for the Advancement of Science. The symposium was on issues of discord between morphological and molecular phylogenies.

Sibley was articulate, rather impressive, and very convinced of the correctness of his technique. He and Ahlquist had just published their analysis of the relationships among the hominoids, and that was the gist of his presentation at the meetings. They had certainly resolved the fuzziness in associating humans and the African apes. They linked the chimpanzee most closely with humans. The gorilla was allied as the sister of the human–chimpanzee group. The orang-utan was the most distantly related among the large hominoids. The gibbon was the sister of the large hominoid assemblage.

A growing number of systematists are understandably accepting wholeheartedly Sibley and Ahlquist's DNA hybridization phylogenies. The idea that the entire genome is being analyzed does appeal. But there is a question that can be raised. And that is about the nature of the difference that one is dealing with and interpreting in a particular evolutionary way.

Only if you assume first that molecular change occurs in a very specific and singular fashion can you conclude that overall genomic similarity — or any overall similarity — equates with phylogenetic closeness. Emile Zuckerkandl and Linus Pauling admitted that such an assumption was necessary in order to understand their data on hemoglobin in a particular evolutionary framework. But similarity and dissimilarity can be interpreted in more than one way.

As I discussed in an earlier chapter, similarity can arise in organisms in two ways. Organisms can be closely related and will thus have inherited similarities from their most recent of common ancestors, or organisms can inherit similar features, unaltered through a chain of predecessors, from a distant ancestor they share broadly with other members of their group. Not all instances of similarity reflect closeness of relatedness.

Dissimilarity can also arise in two different ways. First, an organism can be left behind evolutionarily while the ancestor of its nearest relatives assumes more novelty; in this instance, the organism left behind remains primitive when compared with relatives that have become more derived as a result of their inheriting those evolutionary novelties their ancestor acquired. Second, that organism itself could change radically, with the result that its kin, including its closest relative, are left behind in its evolutionary dust; although they have remained primitive, the relatives that have been left behind would appear more similar overall to one another than any would to the organism that charged ahead evolutionarily. If you assume that *in every situation* greater similarity is a reflection of greater closeness of relatedness, you will interpret some relationships the wrong way around. You will conclude that the taxa left behind are similar because they are closely related to one another and evolved together, and that the organism that actually charged ahead is dissimilar because *it* was the one that was left behind.

With DNA hybridization, you cannot tell the directionality of similarity and dissimilarity. Are organisms similar because they have actually changed and left the dissimilar ones behind? Or is the dissimilar organism the one that changed and left the others behind? These questions are equally applicable to *any* analysis in which you cannot determine the nature of the similarities and the dissimilarities. Immunological and antigenic reactivity studies are similarly problematic. The reaction between antiserum and serum might indicate that difference exists, but it cannot tell you in which ways organisms really are different from each other.

In a column in *Science*, Roger Lewin featured Sibley and Ahlquist's hominoid research. Trying to present alternative points of view, Lewin quoted a dissenting comment from the Washington University evolutionary biologist Alan Templeton. Templeton had, a few years earlier, published his own analysis of hominoid relationships based on DNA specific to the mitochondrial organelles of the cell. And Lewin quoted Templeton as objecting to the use of DNA hybridization because it didn't allow one to determine the polarity of the similarity — is similarity due to distant or recent common ancestry? Templeton argued that only by studying the actual sequences of nucleotides can one determine the significance of similarity and dissimilarity.

Having the actual, identified nucleotide sequence, and being able to compare such sequences broadly among organisms, should give us some necessary insight into the ways in which organisms are similar and different. However, Sibley and Ahlquist maintain that there is "no reason to expect data derived from base sequences to improve on those from amino acid sequences, which have produced contradictory results." Sibley and Ahlquist even quote A. E. Friday (of Romero-Herrera's team) as noting that "phylogenetic conclusions derived from a study of nucleotide sequences will be subject to the same suspicions as those derived from amino acid sequences."

So where are we? If you listen to enough comments from the array of biochemists and molecular systematists, you can easily come to the conclusion that only the one you happen to be reading at the time can be believed; that only the particular technique being espoused at that moment can provide the correct answers to phylogenetic questions. But even in the face of such apparent lack of unanimity, there still remains the tendency to feel that the bulk of the molecular and biochemical data has supported the unity of humans with one or both of the African apes.

To me, there is a major inconsistency.

Molecular change is supposed to occur in the same fashion, across all molecules and all organisms, regardless of the taxa involved. Such change should affect all taxa in similar ways, with the result that molecules and organisms can be studied and analyzed in similar ways. Thus, if there is apparent congruence between different techniques in one part of their resultant phylogenies, there should be congruity throughout the phylogenies. But as I reviewed in the previous chapter — for instance, with Sarich's and Goodman's analyses of prosimians and hominoids — this is hardly the case. How can we accept as unassailable the unity of humans and the African apes when the rest of the phylogeny generated makes no sense, or is incompatible at similar levels of detail with other phylogenies?

Study of myoglobin might demonstrate the greater similarity of humans and the African apes, but what about the "enigma" surrounding the orang-utan? Hemoglobins might appear identical in humans and the two species of chimpanzee, but what about the contradictory theories of the relationship of the tarsier? Humans and the African apes might emerge as similar immunologically in some of their blood serum proteins, but why is there no agreement on the details of the relationships among the prosimians? It is also the case that depending on the particular study of blood serum protein electrophoresis one chooses to accept, the relationships among the hominoids do not even come out as expected.

There have been various reasons given to explain why the results on the hominoids should be embraced and why, all the same, one should expect that the proposed relationships among the prosimians might not be reliable. The essential element is time.

The evolutionary events which produced the hominoids are much more recent than those which resulted in the diversity of prosimian primates. Because of that, the argument goes, there has been less time since the differentiation of the hominoids during which molecular "glitches" could have been introduced. Thus, the results on the hominoids (and any other group of relatively recent origin) is cleaner. In contrast, in the tens of millions of years since the diversification of the prosimians is supposed to have taken place, various molecular changes and flipflops had plenty of time to have occurred and produced false evolutionary footprints. Thus, the older the lineage, the more interference there will be in arriving at the correct resolution of relationships. It is believed that a major source of this interference comes from the transitions that occur from one nucleotide to another.

It appears that there is a tendency for one large nucleotide to be replaced by the other kind of large nucleotide. Similarly, there are transitions between the two kinds of small nucleotides. It is apparently much rarer — and perhaps chemically more difficult — for a large nucleotide to be replaced by a small nucleotide or, conversely, for a small nucleotide to be supplanted by a large nucleotide. The substitution of one major category of nucleotide by another would presumably produce change that would have profound effects. The transition between one kind (let's call it state A) and the other (let's say state B) of the same category of nucleotide would have less profound effects, but it nevertheless would still constitute change.

The problem, however, is that a nucleotide can change from state A to state B and then back to state A again without your knowing that anything has happened. The "change" would go undetected. Or a nucleotide could change from state A to state B to state A and then to state B again, and you might not know that this happened either. You would record one change, not the correct number, three.

Think of the Marx Brothers' film *Duck Soup*, the scene in which Groucho is in a nightgown and nightcap and is chasing Harpo, who is made up to look like Groucho, from the sleeping clothes to the fake mustache. During the chase, a large mirror in the house is broken when Harpo runs into it. But Groucho is close behind, so, in order not to get caught, Harpo has to pretend he is Groucho's reflection in the "mirror."

They go through a bunch of shenanigans, with Harpo following Groucho's lead as he hops and scoots back and forth in front of the "mirror." But at one point Groucho decides to get sneaky. He stands looking at the "mirror" and then spins completely around, ending up facing the "mirror" again, with his arms outspread. Harpo just stands there, facing out of the "mirror," and, at the crucial moment, spreads out his arms. We know that Harpo did not move an inch, but Groucho does not know that. While this "joke" is hilarious to watch, it is also exemplary of problems in biomolecular approaches to evolutionary relationships.

Harpo is analogous to a nucleotide that did not alter its condition. Groucho is one that did — going from state A (in this case, facing forward) to state B (facing backward) and then back to state A again. But if all we could see was Groucho and Harpo facing us in the same forward position, we would never know if anything had happened, or to whom it had happened. Multiply this example by millions of nucleotides and you will get a sense of how much change could have occurred without there ever being a possibility of its being detected. If you do happen to see sequence similarity in DNA, is it really homology and thus really indicative of relatedness, or is it just sequence similarity, with superficially similar sequences being the result of different molecular events?

There is supposed to be a correlation between the amount of time since the divergence of taxa and the amount of accumulated nucleotide change in the emergent taxa. The less time that has elapsed since the diversification of related organisms, the less is the chance that undetected changes in nucleotide identity will have occurred. With more time, however, the possibility is introduced that the similarity or dissimilarity which one demonstrates among organisms will be colored by the "silent" and undetectable transitions among nucleotides. Thus, false phylogenies will become increasingly more likely as the age of the group under investigation becomes more ancient.

There is a way to get around this problem. Some molecules apparently have slow rates of nucleotide substitution, and they can be used to unravel the details of the relationships of evolutionarily old groups. Other molecules, with rapid rates of nucleotide substitution, are preferable for deciphering the relationships of those groups of more recent origin.

Wes Brown, now at the University of Michigan, and others, including Allan Wilson at Berkeley, have worked on the DNA of the mitochondrial elements of the cell. Mitochondria appear to be separate little islands within a cell, which, although now aiding in some cell functions, may have originally been invasive, parasitic elements. Brown and his co-workers concluded that mitochondrial DNA evolves (or at least changes) five to ten times more rapidly than DNA found in the nucleus.

This conclusion was made by observing that there was more difference between humans and a sampling of Old World monkeys in their mitochondrial DNA than there was among these same primates in their nuclear DNA. Assuming that the divergence between the human and Old World monkey lineages occurred over twenty million years ago, the differences between these catarrhines in their nuclear and mitochondrial DNA could be explained if nucleotide substitutions occurred at a slower pace — five to ten times slower — in nuclear than in mitochondrial DNA. In one article, the researchers calculated that mitochondrial DNA data are the most accurate for dealing with evolutionary events that occurred "within the past 3–10 million years." In 1981, using this investigative tool, they approached the relationships of the hominoids, whose evolutionary history supposedly falls reasonably within the limits of the technique. The result of these studies was that humans, chimpanzees, and gorillas were most similar in their mitochondrial DNA and thus most closely related. It could not be determined which two of these three hominoids were the most closely related.

In a 1982 study of hominoid mitochondrial DNAs, Brown and his team favored uniting the African apes even though by their own data and method of phylogenetic reconstruction they should have separated the chimpanzee from the gorilla and united it most closely with humans. A review of their data reveals that there are fewer substitution differences between humans and chimpanzees than between humans and gorillas. If humans, however, are related to an African ape group, then the number of substitution differences should be equal between humans and chimpanzees and between humans and gorillas — this, of course, follows from the basic assumptions about molecular change.

There are a few peculiar aspects to the use of mitochondrial DNA in the reconstruction of evolutionary relationships. One striking feature of mitochondrial DNA is that it is not inherited in the same manner as regular, nuclear DNA. Mitochondrial DNA is inherited only through the maternal line. It is not passed on from the father, unless, by some freak accident, some mitochondrial DNA from the tail of the sperm gets into the egg it fertilizes.

Crucial to the work of Brown and his colleagues is the notion that analyzing the mitochondrial DNA is good only for evolutionary events that occurred within the past three to ten million years. For instance, the nucleotide difference between humans and mice (whose lineages may have diverged about eighty million years ago) "will not be appreciably different from that between species which diverged only 20 million years ago, because the readily-substituted positions in the [mitochondrial] DNA have become 'saturated' by this time." Thus, Brown and colleagues conclude it is important and "necessary to choose a series of species for comparison whose divergence times are (1) different and (2) lie within a time range that is short enough to give a favorable signal-to-noise ratio." In other words, you have to choose species whose evolutionary history falls within the ten-million-year "good" period of mitochondrial DNA. With longer periods of time, there will be more molecular noise, or interference, through the buildup of substitutions that will cloud the real picture.

A corollary to the time limit on mitochondrial DNA is that closeness in time requires researchers to study organisms that are closely related. The extant hominoids fill the bill. They are supposed to be a closely related group of primates which apparently differentiated within the time frame appropriate for mitochondrial DNA analysis. Brown and his colleagues used Sarich and Wilson's earlier work as evidence of the youthfulness of the hominoids — as proof that, indeed, "the most distantly related hominoid species probably diverged from one another no more than 10 million years ago." If the hominoids are this young evolutionarily, then "this group of primates can provide information on [mitochondrial] DNA evolution that is relatively unobscured by multiple substitutions at the same nucleotide position."

There are two things very wrong here.

The first is that by choosing a group of closely related taxa to work with, you are already making a major assumption about the phylogeny of the group whose phylogeny you want to unravel.

The second problem is that since times of divergence can be calculated only after the phylogenetic relationships among organisms have been agreed upon, you must choose between the various available theories of evolutionary events of the "group" you are interested in before studying those events. In order to satisfy the criteria of time and relatedness before being free to use mitochondrial DNA data for phylogenetic reconstruction, you must have prior knowledge of the phylogenetic history of that group.

In addition to assuming that Hominoidea does indeed constitute an evolutionary group, Brown and his colleagues relied on Sarich and Wilson's 1967 calculations for the age of this group of primates. But in the 1960s and 1970s the most favored scenario portrayed an Oligocene origin of Hominoidea, perhaps much earlier than twenty-five million years ago. Not all molecular biologists agreed with Sarich and Wilson's dates for primate diversification. Indeed, Sarich would eventually come to add a few million years on to the dates of divergence he had originally arrived at for the hominoids. Even Sibley and Ahlquist's more recently calculated dates of hominoid divergence extend far beyond the ten-million-year constraint imposed upon mitochondrial DNA data, with the gibbons splitting off more than twenty million years ago and the orang-utan lineage emerging about sixteen million years ago. Only humans and the African apes fall within the ten-million-year marker, which is the approximate time Sibley and Ahlquist postulate for the separation of the gorilla from the human–chimpanzee group.

If the origin of Hominoidea predates ten million years, then mitochondrial DNA data cannot support the unity of this group. If this is the case, the monophyly of Hominoidea — a hypothesis derived from other data — must be assumed. The study of mitochondrial DNA cannot resolve this probable relationship. Similarly, if any of the divergences of extant hominoids occurred earlier than ten million years ago, mitochondrial DNA will yield no reliable information on relationships. This is of special concern because one can *never* be one hundred percent certain about any theory of relatedness, or about the evolutionary events that constitute a group's history.

How does one decide which of the available and competing phylogenetic schemes is "correct"? How does one know that the dates of divergence calculated are accurate? And if one has to make these assumptions beforehand, from where does the validation come that gives a theory of relatedness the power to reject competing theories?

If you must rely on other sources of data for justifying the overall relatedness of groups and specific relationships, a real problem emerges when you declare your answers are correct despite their disagreement with the sources you began with. This is the essence of the "molecules versus morphology" issue.

In the beginning, tests of molecular and biochemical theories of the evolutionary relationships of organisms came from morphologi-

cally derived theories of relatedness. Since at that time there tended
to be broad agreement between these different levels of analysis, there
seemed to be some validation of molecular and biochemical approaches to evolutionary issues. Then, when more and more groups
were scrutinized molecularly, and there arose from time to time lack
of agreement with morphologically based theories of relatedness, the
molecular scheme was taken as the more "correct," since it was more
technologically sophisticated.

Today, molecular systematics is becoming increasingly divorced
from morphology. It is now rather common to see molecular theories
of relatedness that separate quite clearly the most obvious of plausible
morphological relatives. One can find examples of this in Sibley and
Ahlquist's DNA hybridization work on various subgroups of birds
and, with regard to hominoids, in the trend toward uniting the chimpanzee with humans rather than with the uniquely anatomically similar knuckle walker, the gorilla. Once you have chosen a particular
theory of relatedness, you are obliged to explain how similarities that
might otherwise support different theories of relationships are actually
parallel and independently acquired attributes. If chimpanzees are
indeed more closely related to humans than they are to gorillas, the
seemingly unique anatomical features shared by the two African apes
must be explained either as parallelisms or as primitive retentions.
On the face of it, that seems fair if you are not dependent at all on
morphology to support any element of your phylogenetic scheme.
However, if you must invoke morphology in any way, then I cannot
see how you can claim, when there is disagreement between morphologically and molecularly based arrangements of taxa, that morphology is deceptive.

As mentioned above, Brown and his co-workers' mitochondrial
DNA analysis relied on Sarich and Wilson's biochemical claims for
the unity of Hominoidea. The initial validation of Sarich and Wilson's
work came from its general agreement with a morphologically based
notion of Hominoidea. Brown and his colleagues used Sarich and
Wilson's date of ten million years for the approximate origin of extant
hominoids. The dates of divergence arrived at by any molecular or
biochemical study derive from the association of a fossil with an extant
form, and that's a morphologically based theory of relatedness.

How many molecular or biochemical studies begin without any
assumptions about the relatedness — or at least a gross approxima-
tion of the general framework of relationships — of the taxa they are

investigating? Very few. The majority of biomolecular studies do one of two things. They confirm broad schemes of relationships among organisms which have usually been derived from other avenues of investigation, or they rearrange a few taxa whose general close relatedness has already been accepted. Nuttall, Zuckerkandl and Pauling, and others approached the matter of molecular evolution broadly among vertebrates. Sarich and Wilson started with a diverse sampling of primates. Brown and his colleagues and Bruce and Ayala, for instance, worked on the more detailed end of things, trying to sort out the relationships among the hominoids.

One of the most frequently cited studies that is supposed to have demonstrated the closeness of chimpanzees and humans, to the exclusion of the gorilla and especially of the orang-utan, clearly states that the authors had first accepted the premise that the orang-utan was primitive. Although I had read and reread this article, published by Jorge Yunis and Om Prakash in *Science* in 1982, it took a while before I caught this comment.

I had initially found it difficult to accept Yunis and Prakash's conclusions about the relationships of the large hominoids because they had dealt only with these four primates and an Old World monkey. I had cited their effort in my *Nature* paper in a brief discussion of the inadequacies of using small samples, without the background of a broad comparison.

The problem of small samples is not a minor point, yet its implications are all too often misunderstood. For example, there was a multi-authored paper a few years ago in *Science* which set out to study the cross-reactivity of a particular type of cell, the "T" cell, of all hominoids as well as a few New and Old World monkeys. The researchers concluded that their data corroborated the notion that humans and the African apes are evolutionary sisters. However, in a rejoinder published later that year in *Science*, Matt Cartmill pointed out the flaw in the argument:

> The data of Haynes *et al.* are compatible with the now generally accepted notion that orangutans are less closely related to us than chimpanzees and gorillas are; but the support they provide for this notion is negligible, because the sole human T-cell determinant that is absent in orangutans is present in gibbons. The only respect in which the T-cell determinants of African apes resemble those of human beings more closely than do those of orangutans is therefore a primitive retention . . . which cannot argue for monophyly of a human-gorilla-chimpanzee grouping.

Even when the slightly broader comparison is available and could provide a check on the significance of similarities in humans and one or both of the African apes, the pull of a human–African ape scheme is apparently so strong that the possession of the same T-cell determinant by the gibbon is overlooked. Overall similarity is regarded as sufficient.

Because this sentiment is widespread, it is no surprise that studies that come to conclusions about the phylogenetic relationships of taxa tend to feel justified in using small samples. Yunis and Prakash's sample did not even include the gibbon. As is obvious from Cartmill's criticism of the T-cell data, however, it is important to look at least as broadly as the gibbon in order to check the significance of the sharing of any particular feature by humans and one or both of the African apes.

The situation with the chromosomes is even more complicated because Yunis and Prakash assumed first that, of the four large hominoids, "the orangutan is the more primitive species." As I see it, they left very little room for entertaining alternative schemes of relatedness. A phylogenetic scheme had already been assumed. The result was not the generation of a theory of relationships based on the analysis of chromosome data, but an explanation of what chromosomal changes would have had to have occurred in order to arrive at that particular scheme of relatedness.

Although I would have preferred a larger sample of primates to work with, I decided to see what groupings of the four large hominoids would emerge if I linked together taxa solely on the basis of their sharing homologous chromosomes. I made many photocopies of Yunis and Prakash's published diagrams of the banding patterns of the chromosomes of the four large hominoids and then proceeded to compare, set by set, the chromosomes that were supposed to represent the same chromosome in all hominoids. In order to keep better track of who shared what chromosome, I devised a color coding scheme, which I superimposed on Yunis and Prakash's illustration of chromosomal banding patterns.

The reason the banding pattern is so important is that when a chromosome is chemically stained, the concentration of DNA at any given position along a chromosome can be highlighted. Thus, a stained chromosome will show along its length a series of bands of differing thicknesses and intensities. If the pattern of these thick and thin, light and dark bands appears identical on the chromosomes of two or more organisms, one assumes that these chromosomes are homologous, having been retained from one or more common ancestors.

My comparison of the banding patterns of hominoid chromosome 1 revealed identity between this chromosome in the chimpanzee and orang-utan. For chromosome 2 — which is the tricky chromosome, because reduction in chromosome number in humans is supposed to have involved this chromosome — the gorilla and the orang-utan emerged as being the most similar. For chromosome 3, humans, chimpanzees, and gorillas appeared to have homologous chromosomes. I could not find any large stretch of identity among the hominoids in their fourth chromosome, but in the fifth one humans and the orang-utan had identical banding patterns. All the large hominoids were similar in their sixth chromosome. I carried out this comparison through the twenty-second chromosome as well as the two sex chromosomes, X and Y.

Out of all the possible groupings among the hominoids, the couplets human–orang-utan and chimpanzee–gorilla emerged as sharing the greatest number of chromosomes with seemingly identical banding patterns. Humans and the orang-utan have in common chromosomes 5, a particular pattern of 12 and 19, and most of Yq, the lower extent of the Y (male sex) chromosome. The chimpanzee and the gorilla also share four apparently homologous chromosomes — another configuration of 12 and 19, 20, and 21 — but with different banding patterns from those which distinguish chromosomes 12 and 19 in humans and orang-utans.

The finding that humans and orang-utans as well as chimpanzees and gorillas share different versions of both chromosomes 12 and 19 strikes me as particularly interesting. If humans and orang-utans share one configuration of chromosomes 12 and 19, and chimpanzees and gorillas share another, then at least one of these taxonomic couplets has to be sharing a derived pattern of banding for at least one of the chromosomes involved. Perhaps humans and orang-utans share the derived pattern for chromosome 12 and chimpanzees and gorillas share it for chromosome 19, or it could be the other way around. There is also the possibility that humans and orang-utans as well as chimpanzees and gorillas share derived, but differently derived, banding patterns on both chromosomes. It could also be the case that only one pair — human and orang-utan or chimpanzee and gorilla — shares a derived pattern for chromosome 12 and chromosome 19.

Without at least the gibbon for comparison, one cannot begin to get any insight into which banding pattern on any given chromosome is derived for the hominoids. But we are left with the tantalizing hint of support for the unity of humans and orang-utans as well as for chimpanzees and gorillas — a hint that gains an additional boost from another, but much more intense, analysis of chromosomal data.

A rather complicated quantitative approach to interpreting hominoid phylogenetic relationships from chromosome data was published by a physical anthropologist at the University of California at Los Angeles, Larry Mai. From his calculations, Mai came to the conclusion that the chromosome data supported five equally likely and parsimonious theories of relatedness among the four large hominoids.

One arrangement clustered the African apes together, then the orang-utan, and finally humans. This is Huxley's human–great ape assemblage. Another scheme linked the two African apes together, a linkage to which humans were more closely related than was the orang-utan. This is the theory that replaced Huxley's. But the remaining three phylogenies always grouped together humans and the orang-utan most closely, to which were then related the chimpanzee and gorilla. In one of these phylogenies, the chimpanzee emerged as more closely related to the human–orang-utan sister group than the gorilla. In another, it was the other way around. In the third of these schemes, the chimpanzee and the gorilla were a group that shared an ancestor with the human–orang-utan group. It is this theory of relationships that is the most compatible with the morphological data.

Because the orang-utan supposedly retains a largely "conservative" chromosomal picture, it is commonly believed that the orang-utan is the most primitive of the large hominoids. But having some primitiveness is not the same as being entirely primitive. One expects an organism to have retained at least something from its earlier ancestors. In fact, in the cases I investigated, I often discovered that an organism has more features that are primitive retentions than features shared only with its closest relative or even unique to itself. The orang-utan may be relatively more primitive in aspects of its chromosome patterning than the other large hominoids — but the orang-utan does possess chromosomal features that are otherwise found only in humans.

Recently I was speaking with J. Desmond Clark about molecular, biochemical, and chromosomal reconstruction of relationships. Clark, a professor at the University of California at Berkeley, is perhaps the world's most accomplished African prehistoric archeologist, a pioneer in early fossil hominid studies. His status as a grand senior statesman is reinforced by a distinguished white mustache and Vandyke, from behind which come the restrained understatements typical of some British.

Among other things, Clark mentioned that Jerold Lowenstein, a radioimmunologist turned molecular systematist, had asked him for some samples of various *Australopithecus* specimens so that he, Lowenstein, might run these bits of fossils through his test tubes and resolve the matter of early hominid relationships. Desmond said he had refused the request, adding that he would like to see what Lowenstein and fellow molecular systematists and biochemists would do with a set of samples of unknown origin.

This is a very challenging idea: Don't tell anyone what animal a bone or tooth or blood sample comes from and what that animal's potential relationships might be. Just give the experimenter a bunch of unmarked samples. See what he or she can do with them. Of course, you should code the specimens so you will have a record of which animals the specimens come from. If they are hard-tissue samples, you should grind them up first, or dissolve them, or otherwise obliterate their structure while still enabling them to be subjected to the necessary biomolecular analyses. If all the experimenter had to go on was a series of vials of liquids and powders, could even the identity, much less the affinities, of the animals represented be determined — or even approximated?

This is a particularly intriguing question in Lowenstein's case because he has spearheaded a series of papers claiming to have sorted out the affinities of organisms through biochemistry where morphology and anatomy failed. In a short piece in *Nature* a few years back, Lowenstein and his colleagues boldly stated that "overconfidence in comparative anatomy led prominent scientists astray" with the result that "the relations of human beings to other primates are still debated, and neither anatomy nor currently available fossils are likely to settle the problem of 'man's place in Nature.' " As far as Lowenstein and his co-workers were concerned, "such problems should be settled, or at least bounded, by the findings of molecular biology."

In one investigation, Lowenstein and his colleagues were fortunate to get hold of a sample of muscle tissue from a forty-thousand-year-old baby mammoth found frozen in the wastelands of Siberia. They extracted what albumin they could and tested its reactivity with candidates for possible relatedness to the mammoth. They concluded that the mammoth was related equidistantly to the Indian and African elephants. This conclusion was in keeping with the longstanding morphological interpretation of mammoth–elephant relationships.

There has, however, been disagreement on the relationships of the elusive, and perhaps not yet extinct, Tasmanian wolf. Some taxonomists have grouped the Tasmanian wolf with the dasyurids, a diverse group of marsupials from Australia, Tasmania, New Guinea, and nearby islands. As a group, the dasyurids have come to include the tiniest of marsupials, the pouched or marsupial "mice" and "rats," which are insectivorous, and the carnivorous, somewhat catlike or weasellike Tasmanian "devils" and native "cats." The more doglike Tasmanian wolf would be the largest of these Pacific island carnivorous marsupials.

On the other hand, some paleontologists and marsupial specialists are convinced that the Tasmanian wolf is not a dasyurid. They believe instead that the affinities of this marsupial carnivore of the southern Pacific lie with an ancient fossil group of marsupial carnivores from South America, the borhyaenids.

Armed with a sample of dried muscle from a bone of a Tasmanian wolf which was collected in the very late nineteenth century, as well as with two samples of untanned skin from specimens collected early in this century, Lowenstein and his crew proceeded to extract albumin and compared it with the albumin of various potential marsupial relatives. The Tasmanian wolf came out as most closely related to, but equally distant from, the two dasyurids that had been analyzed. Lowenstein and colleagues concluded that "these results should help settle the controversy about [Tasmanian wolf] affinities . . . for the Tasmanian wolf's albumin is that of a very recently derived dasyurid."

Bolstered by this apparent systematic success, Lowenstein, with two different collaborators, turned his attention to another phylogenetic "mess," the true identity of Piltdown Man's jaw and canine. These bones, which were associated with a modern-looking human skull "discovered" in the early 1900s, had always been recognized for what they are — apelike. However, some scholars had thought the jaw more chimpanzeelike, while others, including the perceptive paleoanthropologist Franz Weidenreich, had found comparisons with the orang-utan more compelling. Subsequent to the revelation in 1950 that Piltdown Man was actually a composite, a hoax, the eminent human biologist John Weiner argued that the jaw and canine were those of an orang-utan. He did not eradicate doubt in the minds of all investigators about the "real" identity of these pieces. Lowenstein and his colleagues admit that "Weiner makes a strong case that the jaw and canine tooth are those of an orangutan, but," they remind us, "he must resort to comparisons of the same kind that led Keith, Hooton and many others to conclude that they were like those of a chimpanzee." As Lowenstein and his associates wrote in *Nature* about the Piltdown case, "it was chemistry [i.e., fluorine tests], not anatomy, which led to the proof that the jaw and skull did not belong together."

Lowenstein and his colleagues found that collagen serum–anti-serum reactions were strongest between extracts from the Piltdown jaw and canine and those from an orang-utan. The reactions were variably less strong when tested against samples from a human, a common chimpanzee, a pygmy chimpanzee, a rhesus monkey, an Indian elephant, and a cow.

Out of curiosity, I decided to map out the rest of their collagen reaction data. After all, a human, both species of chimpanzee, and a rhesus monkey had been included in the analysis. I thought one could at least get a sketchy outline of some primate relationships using this set of immunological data.

I first used just the figures that represented "the relative binding of antisera to various collagen species by extracts of Piltdown jaw." Since "maximum binding is taken to be 1," the Piltdown jaw and the modern orang-utan are immediately associated. The jaw and especially its preserved molars look like those of an orang-utan, so let's assume that the jaw really is from an orang-utan. Taking the Piltdown jaw as synonymous with "orang-utan," we want to look for the highest figure of antiserum binding with another taxon. This turns out to be 0.76, which links the common chimpanzee most closely with the orang-utan. Humans, with a binding figure of 0.61, are farther away. The pygmy chimpanzee, at 0.56, is farther away still, and the rhesus, at 0.35, is the most distant.

The figures on the Piltdown canine yield a somewhat different picture. A binding figure of 1 between the Piltdown canine and a sampled orang-utan means "orang-utan." The closest relative of the orang-utan, with a binding figure of 0.66, is *Homo*. The common chimpanzee is a bit farther away, at 0.62. The pygmy chimpanzee is

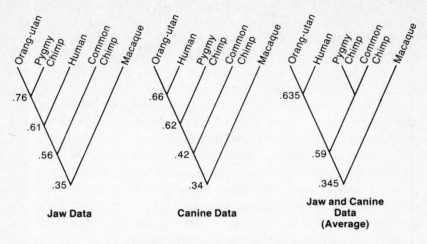

FIGURE 9C   Three different schemes of evolutionary relationships, using the collagen data in Jerold Lowenstein et al.'s analysis of the identity of the Piltdown jaw and canine.

quite distant, at 0.42. At 0.34, the rhesus monkey remains the last.

Since it is common practice in distance analyses to average results, I decided to do that, too, averaging the jaw and the canine figures. Humans came out closest to the orang-utan, at 0.635. The two species of chimpanzee emerged as equally distant from the orang-utan, each having a binding figure of 0.59. The rhesus, again, fell out as the most removed from the orang-utan.

The results surprised me. In the first two "phylogenies," the rhesus monkey emerged definitively as the most immunologically removed, but the other results seemed inconclusive. The two species of chimpanzee were dissociated from each other, as indicated by the fact that, in the jaw as well as the canine comparison, the sample from each species reacted differently to the Piltdown specimen's sample. Only when the results of the jaw and canine analyses were averaged was there reason to associate the two species of chimpanzee. In this case, each species of chimpanzee was found to have a binding figure of 0.59, which, according to the "rules" of similarity, means that these two chimpanzee species must be similar and dissimilar to the orang-utan in the same way. Only through this manipulation of the data do the two species of chimpanzee emerge as intimately related to each other — which is what all systematists, biomolecular and morphological alike, take as a given.

After multiple rereadings of Lowenstein's Piltdown paper, the only way I could see that it differed from the mammoth–Tasmanian wolf paper was that serum–antiserum cross-reactions of samples from living animals were not carried out. Thus, for example, antiserum to common chimpanzee collagen, or to rhesus monkey collagen, was tested only against extracts from the Piltdown jaw and canine, not against samples from any of the living animals. However, if immunologic identity is supposed to reflect the closeness of the Piltdown jaw and canine to the orang-utan, then a natural extension of the approach is that degrees of immunologic dissimilarity should reflect more distant evolutionary relatedness.

Because Lowenstein's technique "permits identification of species-specific proteins," it can be "used to investigate the relations of fossil to living species." Aside from the fact that the mammoth and the Tasmanian wolf are not fossils (at least not yet), the relationships of these "enigmas" were supposedly resolved by the same technique that "unraveled" the mystery of the Piltdown jaw and canine. Thus, if there is a correlation between immunologic similarity and dissimilarity and evolutionary closeness and distance, then the collagen data on the Piltdown bits reflect the orang-utan's closeness to, or distance from, the other primates tested. If, on the other hand, the Piltdown collagen data do *not* lend themselves to phylogenetic reconstruction, then neither do any other studies based on this technique. That, to me, is the logical consequence of the situation.

As I see it, the only indisputable demonstration of similarity that is reflective of evolutionary closeness is identity: "*This* over here is exactly the same as *that* over there." The Piltdown jaw and canine emerged immunologically as belonging to an orang-utan. Not *almost* an orang-utan, or midway between here and there — simply an orang-utan.

A typical comment I have received on my orang-utan papers is that I have totally dismissed the molecular, biochemical, and chromosomal approaches to resolving phylogenetic relationships. To the contrary, I have not. I have, however, tried to understand what makes these analyses tick, and am not convinced that the conclusions reached through these endeavors are necessarily more correct than those generated through alternative means. If there was unanimity among biomolecular analyses at all levels of taxonomic detail, that would be one thing — but there isn't.

All sources of data relevant to the unraveling of relationships should be entertained. When molecules and morphology are in agreement, then maybe the phylogeny under consideration has a good chance of being "correct." When molecules and morphology are not in agreement, it is not necessary to decide which system is "right," it is necessary to ask why they do not agree.

10

The Return
of the Red Ape

I AM SOMETIMES ASKED if I "really believe" my theory on the relatedness of humans and orang-utans. "Belief" is not quite the right way to think about scientific theories, but I guess what it boils down to is asking me if the many months that have passed since I published my original papers on the subject have mellowed my unorthodox view of hominoid evolution. That's not the right way to think about things either. It sounds as if they're about to suggest a long sea voyage — rest up a bit, especially psychologically — after which everything will be all right. Time, however, does not falsify theories.

More than ever I do think there is something viable about the theory that humans and orang-utans are closely related. And that is because at present — and I must emphasize the "at present" — I am not convinced that alternative theories of hominoid relationships are more robust. This does not mean that, over the past months, there have not been some attempts to disprove my ideas. Indeed, Brian Shea of Northwestern University has, for example, argued that ontogenetic changes in craniofacial relationships are distinctly similar in humans and the African apes. But even with this and other recent endeavors, the human–orang-utan theory is too strongly corroborated to be discarded out of hand.

It is an irritating idea because it makes for inconvenience. For instance, if my theory is true, whatever do we do with all the scenarios of human origins based on a chimpanzee model?

If humans and orang-utans are sister taxa, then they had a common ancestor not shared by any other living hominoid. And because we can delineate specific, unique features that humans and orang-utans share, we can attempt a reconstruction of their last common ancestor.

For instance, if humans and orang-utans are united because of their unusual ability to grow the longest hair, or because they among catarrhine primates have the most widely separated mammary glands, or because they have more delayed ossification in forelimb epiphyses than other primates, then the last common ancestor of humans and orang-utans would have had these features, too. This hominoid would also have been characterized by having the longest gestation period among the primates, the highest excretion levels of the sex steroid estriol, and the most pronounced brain asymmetries. Among other anatomical bits and pieces, this human–orang-utan ancestor would presumably have had poorly developed ischial callosities, a single incisive foramen in the palate, a short and deep scapula, low-cusped cheek teeth, and thick molar enamel.

Thick molar enamel? Is that still valid?

It was the issue of the phylogenetic significance of thick molar enamel among the hominoids that got me interested in hominoid systematics in the first place. Thick molar enamel was basically what made *Ramapithecus* (and, later, it was also discovered, *Sivapithecus* and *Gigantopithecus*) related to proper hominids. But thick molar enamel was also found in the orang-utan. However, if the African apes are more closely related to hominids, then the more distantly related orang-utan must have developed its thick molar enamel independently of the ancestral hominid's acquisition of it.

With the recent discoveries of more completely preserved specimens of *Sivapithecus* and the realization that this "dental" hominid had details of the facial skeleton that were otherwise characteristic of the living orang-utan, the phylogenetic significance of thick molar enamel was dismissed. But it need not be, if hominids, the orang-utan, and *Sivapithecus* are closely related.

The effort to dismiss thick molar enamel as significant has begun again, with renewed ferocity. Lawrence Martin of SUNY Stony Brook, for instance, has crammed years of dissertation research into one recent paper for *Nature* entitled "Significance of Enamel Thickness in Hominoid Evolution." Martin was displeased with the previous approaches to measuring enamel thickness. He felt that there was more to having thick or thin molar enamel than just having a thick or thin layer of this hard dental tissue. First, he devised what he

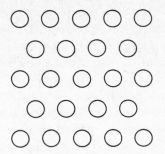

Enamel Prism Pattern 1　　　　**Enamel Prism Pattern 3**

FIGURE IOA The two enamel prism patterns seen in hominoid molars. Enamel prism pattern 3 is commonly referred to as the "keyhole" pattern.

thought was a more accurate technique for determining enamel thickness. Then he found that the chimpanzee, the gorilla, and the gibbon had relatively thin enamel whereas *Homo* and *Sivapithecus* had relatively thick enamel. The orang-utan's enamel was thicker than in the African apes and the gibbon, but not quite as thick as in humans and the orang-utan-like fossil *Sivapithecus*.

Next, Martin examined the microstructure of the various enamels and found differences among the hominoids in the way the microscopic units of enamel, prisms, are packed together. The two patterns of enamel prisms had previously been classified as "pattern 1" and "pattern 3." Pattern 1 prisms are fairly circular in cross section and arranged in discrete rows like walls made from the bottoms of soda bottles.

Martin found that pattern 1 enamel is found in all hominoids. There is one layer of it just over the softer dentine of the inner structure of the tooth, and this layer marks the onset of the enamel. The second layer of pattern 1 enamel is found on the tooth's surface, and its presence reflects the termination of enamel production.

Enamel prism pattern 3 has been nicknamed the keyhole pattern: the stem of one keyhole is sandwiched between the roundish top part of the two keyholes below it. The general effect is of alternating rows of prisms, with each row locked into place by its stems being caught between the heads of the keyholes of the row below. Martin found that pattern 3 enamel is deposited about three times faster than pattern 1 enamel.

The reason "gibbons have thin enamel," Martin wrote, is that "it forms for a relatively shorter period (in relation to size) than does enamel in species of the great ape and human clade, which have extended developmental periods for tooth enamel relative to size." So, another factor in the quality of thick or thin molar enamel is how long there is for enamel to form. Gibbons have enamel prism pattern 3, which is a fast-forming enamel, because the period of time over which their enamel is formed is rather short. Humans have thick enamel because there is more time available for enamel to form. Theoretically, I suppose, if one could somehow truncate the human gestation period and juvenile phase of development, humans, like gibbons, would have only a thin layer of enamel.

The chimpanzee and gorilla should have thick molar enamel because they have longer developmental periods than the gibbon. However, they do not. Chimpanzees and gorillas have thin molar enamel, but their thin molar enamel is not the same as the thin molar enamel in the gibbon. A substantial portion of the molar enamel of the chimpanzee and the gorilla is prism pattern 1 enamel, which is enamel that has a slow rate of deposition. Thus, according to Martin, the thin enamel of the chimpanzee and the gorilla is better defined as "thin/slowed" in contrast to the "thin/fast" enamel of the gibbon.

The orang-utan has relatively thicker enamel than the gibbon and the African apes, but its enamel is not quite as thick as in humans. The reason is that the surface layer of pattern 1 enamel is relatively thicker in the orang-utan. Thus, there may have been a slowing down of enamel deposition in the orang-utan, but a slowing down that was not as drastic as it was in the African apes.

Martin then proceeded to interpret the "significance of enamel thickness in hominoid evolution" by superimposing his data on a preconceived scheme of relationships among hominoids, which posited the gibbon as the sister of a large hominoid group in which the orang-utan is the sister of a human–African ape group. Martin justified this particular theory of hominoid relationships by asserting that "the relationships among the living members of the great ape and human clade are based on cladistic interpretation of overall morphological pattern." He was aware of my suggestion of human–orang-utan relatedness, but dismissed it by claiming, "The recently proposed close relationship between human and orang-utan depends very largely on the interpretation of thickened enamel as a shared derived character between these two genera, but as this has not been shown to be the case, and as our main aim here is to determine ancestral conditions, this view is not accepted here."

By accepting a human–African ape theory of relatedness, Martin can proceed to make various interpretations of the significance of enamel thickness among the hominoids. The common ancestor of the gibbon and the large hominoids would have had thin/fast enamel. The common ancestor of the four large hominoids would have had thick/fast enamel. The orang-utan would then have departed from this ancestral large hominoid condition by developing intermediate/thick/slowed enamel. The last common ancestor of the African apes and humans would have retained the condition of thick/fast enamel from the large hominoid ancestor, which is the reason that humans have thick/fast enamel. The reason the chimpanzee and the gorilla have thin/slowed enamel is because their last common ancestor actually departed from the primitive large hominoid condition of thick/fast enamel.

Martin studied the microstructure of the molar enamel of *Sivapithecus* and found that it was thick/fast, as it is in humans. Because Martin accepted the more recent suggestion that *Sivapithecus* and the orang-utan are closely related, he concluded that *Sivapithecus*, like humans, had retained its development of thick/fast enamel from the common ancestor of the large hominoids.

The upshot of Martin's efforts is that "thick pattern 3 enamel does not define a hominid." Having hominidlike teeth is not restricted to hominids, *Sivapithecus*, and orang-utans. As far as Martin is concerned, "the common ancestor of the great apes and man, and of the African apes and man, would have had teeth resembling those of hominids."

There are two major things wrong here.

One critical problem is that Martin did not demonstrate that thick/fast molar enamel was primitive for the large hominoids. This conclusion results only as a consequence of assuming a particular phylogenetic scheme. If humans are more closely related to the African apes than they are to the orang-utan, then, of course, thick/fast enamel in *Sivapithecus* and humans could be a primitive retention. However, this is only one possible interpretation. Thick/fast enamel could also be a parallel development in *Sivapithecus* and hominids. If this suggestion is accepted, then we can postulate the condition of thin/fast enamel — which is the condition that Martin suggests is primitive anyway — as having been retained in the ancestor of the large hominoids. Even if you assume that humans and the African apes are sisters, you don't necessarily have to conclude that thick/fast enamel characterized the last common ancestor of the large hominoids.

My second objection focuses on the manner in which Martin dismissed the theory of human–orang-utan relatedness. His stated reason was that this theory "depends very largely on the interpretation of thickened enamel as a shared derived character between these two genera," and this is what his analysis purportedly found not to be the case. But, of course, his "finding" was dependent on the human–African ape scheme of relatedness being correct. And although the thick molar enamel issue was one of the initial reasons I became interested in the problem of hominoid phylogeny, the theory of human–orang-utan relatedness by no means lives or dies by teeth alone.

On the face of it, the most parsimonious conclusion from Martin's data is that the African apes and the orang-utan are united because they have more pattern 1 enamel deposited on their molars than do gibbons or humans; slowed enamel deposition would support Huxley's great ape group. But *Sivapithecus* does not have a dominant layer of slow pattern 1 enamel — it has thick/fast pattern 3 enamel. And *Sivapithecus* is best interpreted as closely related to the orang-utan. Thus, it would seem that the layer of slow pattern 1 enamel in orang-utan molars and in the molars of African apes developed in parallel. Perhaps, then, the significance of a layer of slow pattern 1 enamel lies in its uniting the chimpanzee and the gorilla as sisters.

At this point, there are two possible routes to follow. One, which Martin favored, is to hypothesize that thick/fast enamel characterized the last common ancestor of the large hominoids, including *Sivapithecus*. An alternative, however, is to suggest that while having pattern 3 enamel in and of itself might not be significant — because the gibbon has it — having *thick* pattern 3 enamel, as in *Sivapithecus* and *Homo*, and even somewhat in the orang-utan, is significant. Pattern 3 enamel itself might be a more common pattern, but having a thick layer of it is not. Thus, *Sivapithecus* and *Homo*, at least, would be united by having thick enamel. The orang-utan can also be included in the human–*Sivapithecus* clade because it has a relatively thicker layer of enamel than the gibbon and either African ape, and because it appears to be most closely related among all known hominoids to *Sivapithecus*.

Enamel thickness may support the unity of the African apes as a sister group, but it does not lend itself as a primary characteristic to resolving the relationships of the African apes within Hominoidea.

The impact of Martin's work has already been felt. In the short time since its publication, I have heard and seen it cited as definitively debunking the usefulness of enamel thickness in sorting out relationships among the large hominoids. In a recent issue of the *Leakey Foundation Newsletter*, Martin continues to claim that his data lead to the rejection of the human–orang-utan theory. His work has even been singled out in the news section of *Science* for the reason that the enamel data support the grouping of humans closely with the African apes, which is contrary to the favorite human–chimpanzee theory of some molecular systematists. But, of course, the enamel data do no such thing. The enamel data only lend primary support to the possible unity of the African apes — whatever the other relationships within Hominoidea might be.

There is a tendency among systematists and their followers to discard a theory if they think they have found fault with one part of it. If I had relied almost entirely on the presence of thick molar enamel for suggesting that humans and the orang-utan were closely related, then, if it were argued convincingly that the thick molar enamel of humans was not homologous with the thick molar enamel of orang-utans, there would be a very good case for rejecting my theory. However, if we eliminate thick molar enamel from consideration (for the moment or entirely), the theory that humans and the orang-utan are closely related is still supported by a diversity of other features.

Another tendency among systematists is to feel that they have demonstrated cause for rejecting a theory by providing apparent support for another, competing theory. A paper given at the most recent meetings of the American Association of Physical Anthropologists proceeded in just this manner.

M. Ashraf Aziz and Samuel Dunlap, two anatomists at Howard University, have had the opportunity to dissect a few specimens of each of the large hominoids and have also collected anatomical data on a few New and Old World monkeys. Their presentation at the meetings sought to demonstrate that, indeed, humans and one or both of the African apes share apparently derived limb muscle configurations, to the exclusion of the orang-utan. The addition of these few anatomical features to those that surfaced from the literature meant, Aziz and Dunlap concluded, that humans were more closely related to the African apes than to the orang-utan. They also took me to task for ignoring the anatomical data that, obviously, supported a theory of hominoid relationships different from the one I favored.

A doctoral student of mine, Linda Winkler, whose research topic is the head and neck anatomy of the orang-utan, had been finding that variation in muscle number and arrangement was rather common. I had concentrated my studies on what I thought would be more solid comparative ground. Indeed, as I remembered from the human and comparative anatomy courses I took as a student, it is common knowledge that variation in muscle patterns can complicate the picture. It had seemed to me while I was initially pursuing the comparative literature that it would hardly do for me to cite data which might be found to be less reliable because of a known tendency to vary.

But even if I had ignored the available literature on limb musculature because it did not seem to support my hypothesis, and given that Aziz and Dunlap have shown that humans share uniquely with one or the other of the African apes some features of limb musculature, does that falsify the theory that humans and the orang-utan are closely related? No. Of course not. There may be some additional support for the suggestion that humans and the African apes are related, but it does not falsify the competing theory. Only if the resultant corroboration of a human–African ape scheme were greatly to outweigh the support for any other competing theory of large hominoid relatedness would the alternatives be considered falsified.

The problem with using the information on limb musculature in support of a human–African ape arrangement is that it appears that humans are more similar to the gorilla than they are to the chimpanzee. Thus, it would be inappropriate, as Gregory did earlier with similarly skewed data, to claim that humans are closely related to *both* of the African apes. Indeed, as John Fleagle and his colleagues from Stony Brook demonstrated at the same meetings of physical anthropologists, if you take hundreds of osteological and muscular features and link the hominoids solely on the basis of overall similarity, humans and the gorilla emerge as the most similar living hominoids. But, then, we have known this for more than a century, since Huxley's *Man's Place in Nature*.

Perhaps the largest flaw in systematic studies is the lack of unanimity — even a truly common language — on the most basic of theoretical and methodological principles. We might all agree that evolution
"exists," that it is the cause of the diversity of life on earth as we
know it. We might even come to some general agreement on the
major subgroupings of life forms, and, perhaps, even on smaller and
smaller segments of these larger groups. But congenial communication
eventually breaks down. Morphologists and paleontologists come to
disagree among themselves. Molecular systematists come to disagree
among themselves. Ultimately, at times, it seems as if no one is speaking to anyone else.

I think there is a theory applicable to both the morphological and
the molecular. I do not think it is an "either/or" situation. And I think,
with regard to the molecular picture, that Britten and Kohne and
subsequently King and Wilson gave us large parts of the answer.

A little discussed aspect of Britten and Kohne's paper in *Science*,
of almost twenty years ago now, is the mechanism they proposed for
the introduction of new families of nucleotide sequences into the
genome. Usually one sees Britten and Kohne's paper cited for being
one of the first analyses using DNA hybridization, or for its demonstration that large segments of the genome are composed of repeated, apparently inert, DNA sequences. But Britten and Kohne were
also greatly concerned with how molecular change might be effected.

True, Britten and Kohne reiterated the common assumption about
changes occurring in nucleotide sequences slowly over time. But they
concluded that this model of change would not be sufficient to explain
the introduction of new families of nucleotide sequences. This, the
introduction of novelty, Britten and Kohne hypothesized, would
"result from relatively sudden events," which "might have profound . . . results on the course of evolution."

How can one test this suggestion? There is a school of biological
thought that believes that anything worthwhile must be testable in
the lab, by experimental design. But with most things evolutionary —
particularly because we are dealing with a past, a history that has
happened once and in one specific way — tests by experiment are
definitely inappropriate. You would have to test this suggestion against
an assumed phylogeny, in a manner similar to testing the validity of
homology. In this way you could see if there were any molecular
oddballs among the more placidly related kin. But if the way in which
you generate a phylogeny is by associating taxa on the basis of those

which are the most similar, then you will never be able to identify, much less admit the possibility of, a "pulse" or sudden shift within the low roar of continuous, concerted change.

Since the justification for assuming that molecular change occurs in rather continuous fashion came from finding apparent agreement between broad-scale aspects of morphologically derived phylogenies, then we might very well be allowed to ask other questions of the molecular data when aspects of those phylogenies are inconsistent with morphologically based phylogenies. For example, comparative anatomy, development, and physiology support the suggestion that humans are more closely related to the orang-utan than to any other extant primate. How do the molecules stack up? As a handy and readily available reference, let us use the summary of molecular comparisons in Peter Andrews and Jack Cronin's review article in *Nature*.

In comparisons of amino acid differences in the fibrinopeptide protein sequence, humans, the chimpanzee, and the gorilla emerge as having identical sequences; humans differ from the orang-utan by two amino acids, as do the chimpanzee and the gorilla. In percentage of nucleotide pair mismatch, as determined by DNA hybridization, humans and the African apes differ by 1.1%; the African apes differ from the orang-utan by 2.0%; and humans differ from the orang-utan by 2.4%. I think these two examples are representative of many other studies.

If humans are more closely related to the orang-utan than they are to either or both of the African apes, then we have to explain why the orang-utan differs from humans in its fibrinopeptide protein sequence by two amino acids, whereas in humans and the African apes the amino acid sequences are identical. We also have to explain why, in its DNA, the orang-utan differs from *Homo* by just over 1% more than the African apes do. My answer is that the orang-utan changed more from the ancestral genomic condition of the large hominoids than the African apes and humans did.

I suggest there is a correlation between the possibility that the orang-utan is genomically more novel than the other large hominoids and the fact that morphologically the orang-utan is strikingly different from the other large hominoids. For instance, think about the orang-utan's large cheek flanges, its development of a throat sac, its lack of a ligament securing the hip joint, the extreme reduction (if not almost loss) of the thumb and big toe and their associated nails, and the extensive wrinkling of the enamel on the molars and premolars. All of these features distinguish the orang-utan from the other hominoids and from primates in general. Steve Ward has also recently pointed out that the orang-utan appears to be unique among the hominoids (and, presumably, mammals in general) because it has only nerves, not both nerves and arteries, penetrating its incisive foramen. At least Peter Andrews would like to interpret this to mean that the single incisive foramen of the orang-utan is not homologous with the single incisive foramen of hominids, but this, I think, would be inaccurate. The discovery of only nerves in this region in the orang-utan is instead yet another indication of the numerous ways in which the orang-utan is distinguished among the extant large hominoids, and does not contradict the evolutionary significance of humans and orang-utans both having a single incisive foramen.

We have here an animal — the orang-utan — that is in many ways more unlike the other large hominoids than they are unlike one another. And, also in contrast to the other large hominoids, we have an animal — again, the orang-utan — that is slightly more incongruous than its closest living relatives in its genomic makeup.

Perhaps the reasons for the orang-utan's morphologic and genomic singularity among the large hominoids are interrelated. Perhaps the genomic "difference" we are picking up between the orang-utan and the other large hominoids is directly related to, and the cause behind, the extensive and obvious ways in which the orang-utan differs morphologically from the other large hominoids.

What does it mean, in terms of any integrative evolutionary theory, that the orang-utan differs from the other large hominoids by two amino acids in its fibrinopeptide sequence, or by just over 1% in its DNA?

We know that amino acid sequences are derived from DNA nucleotide sequences. Therefore, if the orang-utan is different from the other large hominoids in some of its protein sequences because it has gone away from the primitive conditions that the others have retained, then the relevant parts of the orang-utan's DNA will be different from that of the other large hominoids for similar reasons. But that is essentially the extent of our direct connections to the heart of the genome, DNA.

The fact that the genetic distances, or overall genomic differences, between any of the large hominoids seem disproportionately small compared to the apparent morphological differences among these primates relates to the point King and Wilson were trying to make about changes in regulatory genes. Small molecular changes in a regulatory gene could have a larger, more profound effect on the resultant morphology through a cascading or "snowball" effect on the timing of the structural genes it regulates. Given a few molecular alterations — the introduction of novel sequences perhaps rather suddenly — genomic difference, and its consequent morphologic difference, could result. Stephen Jay Gould and Niles Eldredge have even suggested that changes in regulatory genes might play an important role in the origin of new species.

Bruce and Ayala, in their electrophoretic analysis of blood serum proteins, discussed King and Wilson's invocation of changes in regulatory genes. They concluded that "this is an intriguing, albeit as yet far from substantiated, hypothesis." But the substantiation of any hypothesis other than one that gives back the same answer of constant change has to be able to test both the gradual and "sudden" models of change. Otherwise, you get out only what you put in — and that is circular.

I think morphologists have unwittingly aided in diminishing the credibility of their analyses in resolving issues of evolutionary relatedness. For decades, morphologists — and I am including paleontologists — have tended to point out with great relish how many times such-and-such a feature has evolved independently in various different lineages of animals. The traditional functional morphologists have made careers out of demonstrating how one can manipulate the shape of a muscle or bone by changing the diet of experimental animals, or the temperature, noise levels, or other aspects of the surrounding environment of their caged subjects.

If morphology is so plastic and pliant, and so responsive to every whim and caprice of which natural selection (or anything else) is capable, how could you ever expect to declare with any certainty that any demonstrated similarity between organisms is also a demonstration of homology? And if you cannot demonstrate homology with any certainty, then you certainly cannot generate theories of relatedness. But we *do* have the apparent certainty of homology — or at least sequence similarity — through modern chemistry. At first glance it does seem true that "the gene's where it's at."

Let us assume that morphology is a fair-weather friend of phylogeny, that morphology can change unreservedly. Where does morphology come from, anyway? Obviously, it derives from some parts of the genome. And if there are some parts of the genome that are capable of directing morphologic change, then there are levels of the genome that must also behave in ways other than the gradual and expected. Bear in mind that, theoretically, at least, these genomic "shifts" do not have to be huge. They might register only as minor differences in genetic "distance" or overall genomic dissimilarity between taxa that look — morphologically — worlds apart.

It has been suggested by some molecular biologists that only about one percent of the genome is expressed morphologically. But as far as virtually all organisms are concerned, we know barely one percent of this one percent about the nature of the "link" between any morphology and its underlying genetic base. When we measure overall genomic similarity, how much of the overlap is in the morphological side and how much is in the "silent" component of the genome?

A lot of activity these days is directed toward resolving, through molecular analysis, the issue of whether the chimpanzee is more closely related to humans or to the gorilla. As seen in Sibley and Ahlquist's DNA hybridization study, for example, the chimpanzee is emerging as genomically more similar to humans than it is to the gorilla — which is surprising because, as far as all morphological studies have demonstrated, the chimpanzee and the gorilla are, essentially, small and large versions of the same animal. But even if this morphological identity is not so precise, the chimpanzee and the gorilla are certainly more similar to each other in specific and unique features than either of them is to *Homo*. Yet the gorilla appears to be genomically less similar to the chimpanzee than the chimpanzee is to *Homo*.

In the case of the orang-utan, the genomic dissimilarity that results in the separation of this animal from the other large hominoids is seemingly consistent with the observation that this animal is also morphologically set apart from the other large hominoids. However, the typical interpretation of molecular and morphological dissimilarity — especially dissimilarity relative to modern humans — has resulted in the slighting of an extremely viable theory of hominoid relatedness: the theory that humans and the orang-utan are sister taxa.

In the case of the gorilla, interpretation of genomic dissimilarity leads to the separation of two taxa (the chimpanzee and the gorilla) that are obviously more morphologically related than the two taxa (the chimpanzee and *Homo*) associated because of their overall genomic similarity.

The ambition to develop the most accurate molecular approach to phylogenetic reconstruction — in both the association of taxa and the determination of the dates of divergence of taxa — is epitomized, I think, in a series of papers headed by the Japanese statistician and mathematician Masami Hasegawa. Hasegawa and his colleagues worked with mitochondrial DNA data and not only apparently corroborated the association of the chimpanzee with humans rather than with the gorilla, but, in their latest endeavor, estimated the split between the chimpanzee and humans to have occurred approximately 2.5 million years ago. They also calculated the divergence of the gorilla at approximately 3.6 million years ago. The orang-utan is supposed to have split off much earlier, about 11.6 million years ago.

Various *Australopithecus*, most notably Johanson's *Australopithecus afarensis*, were extant more than 2.5 million years ago. In fact, it appears that *afarensis* and perhaps the hyper-robust species of *Australopithecus* existed more than three million years ago, with some specimens attributed to *afarensis* being closer to four million years in age.

If the chimpanzee is the extant sister of *Homo* and the split between these two lineages occurred approximately 2.5 million years ago, then the taxon *Australopithecus* is rendered asunder. *Australopithecus africanus*, *Australopithecus robustus*, and *Australopithecus boisei* are found on either side of this hominid–chimpanzee split. Forget about *Australopithecus afarensis*; it has just been severed from further consideration.

Perhaps one could say that these "species" never did belong to the same genus, anyway, so nothing is really being violated. But when morphologically relatable specimens attributed to the same species of hominid — even if the resultant hominid species do not belong to the same genus — are scattered on either side of a postulated common human–chimpanzee ancestor, then I think it is time to stop and take stock of what has been going on. Even dyed-in-the-wool molecular systematists like Vince Sarich are compelled to admit the viability of *Australopithecus*.

It is one thing to pull apart morphologic kin such as the chimpanzee and the gorilla, which, admittedly, are not identical down to the letter.

But it is another to suggest that there is nothing at all in morphological similarity which has any phylogenetic credence — unless you are willing to take all of the consequences.

If you adhere to the logical outcome of what are probably very state-of-the-art statistical techniques for dealing with the complexities of DNA sequences, then in my opinion you are the architect of the dissolution not only of morphological systematics, but also of molecular systematics. If morphology is phylogenetically so deceptive that two specimens which would be attributed to the same species cannot be, then none of the morphologically based phylogenies that originally gave validity to the use of molecular approaches to phylogenetic reconstruction, and none of the dates of divergence derived from the association of fossils with extant taxa, are reliable. And if calculations of nucleotide substitution rates and dates of divergence are dependent at all on chronological information from the fossil record, then these calculations must themselves be suspect. What paleontologists must realize is that if they subscribe to a biomolecular theory of relationships among extant taxa that is even in part contradictory to what a study of morphology would conclude, they cannot then proceed to add fossils to the accepted phylogenetic scheme because they have already embraced the notion that morphology is unreliable in assessing evolutionary relationships.

As for me, I will take my chances with a head-on approach to morphology and molecules: each reflects phylogeny, and, I think, the same theoretical and methodological concerns should be brought to bear on each arena of inquiry. Perhaps there are things evolutionary that are a bit more specific to the genome–morphology "linkage" than to the families of repeated but inactive DNA sequences. But it is probably incorrect, if not counterproductive, to proceed as if there were two separate evolving worlds, neither accountable to the other.

To discard all morphological features from phylogenetic consideration because some elements might be characterized by their variability and responsiveness to external factors is shortsighted. We do know, for example, that muscle exertion can have an effect on the area of bone to which a muscle is attached. If you chewed rocks for a living, the appropriate jaw muscles would enlarge and distend the bone at their sites of attachment much more markedly than they would if you sat around and consumed mush.

For such major systems as your teeth, the elements of your limbs, and the hard- and soft-tissue components of your skull and face, final shapes and dispositions are determined early in embryonic development. They are prepatterned. There are what are termed "stem" cells for each system, which carry all the genetic information necessary for the development of that system.

By the end of the first fetal month, it appears that all the stem cells which will be responsible for forming a complete, functionally integrated organism are in place. The embryo has limb development under way; soft internal organs and parts of the skeleton have begun to differentiate. And this embryo, which is still very far from being a recognizable human, even has sex cells.

In the fully formed, adult organism, what we see as separate, apparently discrete body parts, each with its own characteristic morphology, are really not such independent entities. A mouse, for example, may have three separate molar teeth, but these teeth are not separate units that were created from independent sources. Andrew Lumsden, who teaches developmental anatomy at Guy's Hospital Medical School in London, has demonstrated that the cells that give rise to the mouse's third molar derive from the cells that formed the second molar, which derived from the cells that produced the first molar.

As Jeff Osborn has put it, the molar teeth are formed by "cloning" from an original stem cell mass, which was the parental cell population for the entire molar tooth class. Although this might not be a popular way to discuss the process (or perhaps it sounds *too* popular, like science fiction), the "clone model" of tooth development makes developmental sense. It also makes sense in the broader biological picture because we know that other systems — like a limb, for instance — result from the proliferation and differentiation of cells that derive from a single stem or parental population of "original" cells. In fact, it even appears that the many different kinds of blood cells are the products of cellular differentiation from an original stem blood cell.

On top of this prepatterning, however, is some plasticity — for example, in the ability to develop larger muscles and more marked muscle markings on the associated bones, or in the ability to live in high altitudes by producing a greater number of oxygen-carrying red blood cells. But these sources of potential morphological variability hardly make you a different animal or species.

It seems that we can view the molecular side of an organism in the same way we can its morphology. There are molecular realms in which there is a notable amount of variability. Among the more obvious in humans are the different blood groups — A, B, and O are the best known — but there is also variability in hemoglobins and other blood serum proteins, and even in mitochondrial DNA. In fact, Wes Brown and his colleagues have published extensively on the unexpected amount of populational variation there is in *Homo sapiens* in mitochondrial DNA. Since comparisons on the molecular level for the purposes of gaining phylogenetic insights usually derive from samples of one or at most a few individuals, it will be interesting to see what happens in the future to the calculation of "genetic distances" when the sample sizes are increased. It is often the case that variation within species is greater than the amount of difference between species.

Thinking in terms of prepatterning, and, thus, the integrity of anatomical systems, puts a different slant on morphology. There is something more compelling theoretically about a few stem cells carrying the genetic information that will orchestrate the unfolding of an anatomical system than thinking of the complexity of an adult individual and trying to figure out the link between all that morphology and the underlying DNA messages. It is not so spectacular and important that my first molar tooth is matched by the first molar tooth of, let's say, an orang-utan. It is significant that entire chunks of our dentitions are similar in ways that distinguish us from chimpanzees and gorillas.

If the molars of humans and orang-utans are distinctive among primates — and this includes the molar teeth as well as the premolars and those deciduous molars which are shed as the premolars replace them — then there must be something about the stem cells of the molar class which distinguishes humans and orang-utans from the other hominoids. And, also because of its molar class teeth, we must include *Sivapithecus* in this discussion. Chimpanzees and gorillas would be distinguished by their own dental uniquenesses, which in turn would be attributable to distinctions in their molar class stem cells.

Since limb bud differentiation parallels what we see in tooth class differentiation, we can perhaps better appreciate the data on forelimb ossification that Schultz gathered for the hominoids. All the hominoids are distinguished from the other catarrhines in their delaying of the onset of ossification of certain of the ends of the arm bones. The large hominoids — the chimpanzee, the gorilla, humans, and the orang-utan — are themselves distinguished by delayed ossification in other epiphyses. Among the large hominoids, humans and the orang-utan are further set apart by their delaying the onset of ossification of yet other forelimb epiphyses. It is not inappropriate to think of these developmental differences as resulting from changes in the messages carried by the stem cells that will guide the course of development of that limb.

Thinking in terms of "developmental packages" instead of discrete anatomical or physiological units expands our appreciation of the significance of gestation length, or high levels of estriol excretion, or the fetal growth zone, encephalization, and development of cerebral asymmetries. The large hominoids are set apart among primates by having longer gestation periods than the rest, some estriol excretion, and more asymmetry of the cerebral hemispheres and Sylvian sulcus. But humans and orang-utans have the longest gestation periods and the highest levels of estriol excretion with concomitantly high fetal growth zones — and these latter features are related to the development of encephalization. As a package, then, these features are certainly of interest when you consider that humans and orang-utans also have the most marked cerebral and Sylvian sulcus asymmetries among primates.

If morphology has suffered at the hands of morphologists who insist that it is subject to any whim of adaptation, so has behavior, for the very same reason.

Sociobiologists have been particularly aware of the difference between attributes of an organism that are invariant and those that are quick to respond to more immediate conditions. In the traditional functionalists' heyday the distinction was, respectively, between features due to adaptation and features that are the result of acclimatization. These days, the more invariant elements — those attributes by which an organism finds itself constrained — are called "obligate" traits, while the more responsive aspects of an organism are the "facultative" traits.

Steve Gaulin, a sociobiologist in my department at Pittsburgh, as well as other sociobiologists, have argued that an animal's diet is determined by its size. Thus, a small animal finds it "obligatory" to seek out concentrated, high-quality food. In contrast, a large animal will characteristically forage over a wider range of diverse, low-quality food sources. The size/diet differences hold up even in highly sexually dimorphic species: the smaller sex will have a somewhat different dietary regime than the larger sex.

Certain aspects of sexual and reproductive behavior have also been perceived as obligate. And size, again, is considered a dominant feature. For example, comparative studies of animals reveal that there is a correlation between the degree of sexual dimorphism characteristic of a species and the mating system that species has. When males are larger than females, even by a small amount, the preferred mating system is polygyny, which is the association of one male to more than one female. The only truly monogamous animals are those in which there is no sexual dimorphism, such as gibbons. Polyandry, where there is one female to more than one male, is correlated with the female's being larger than the male.

Although we in the Western world like to think of humans as monogamous, the tendency worldwide is toward polygyny, and it is the case that human males are, on the average, a bit larger than females. Orang-utans, among the most sexually dimorphic of primates, travel in such dispersed fashion that they appear to be solitary animals, but they actually have a loose "group" composed of an adult male and more than one adult female. This type of male/female ratio is quickly noted in the small, closely associated group typical of the gorilla. Male access to more than one female is most extreme in the frenetic and promiscuous chimpanzee.

Such attributes as the length of a copulation bout and the development of estrus have been interpreted as facultative in their expression. I have not seen, however, any argument for why the time spent in copulating is the longest in humans and orang-utans (orang-utans may even have longer copulatory bouts). But there is an inclination among some sociobiologists to attribute the development of estrus, or the lack of it, to the particular needs of a species.

For instance, humans and orang-utans are unique among the large hominoids — and probably among primates, depending on what future studies on gibbons and a few monkeys yield — in not developing any female sexual swelling coincident with the peak of the ovulatory cycle. There appears to be a concentration of copulatory activity around the peak of ovulation, but female humans and female orang-utans have the potential for sexual receptivity during the entire menstrual cycle — and in this they are definitely unique.

It has been explained to me that although humans and orang-utans appear so physiologically and behaviorally similar, they arrived at this state for different reasons. A common scenario for humans is that the female genital region became hidden with the evolution of upright posture and bipedalism. Why go to all the trouble of trying to signal a male with genital swelling if no one can see it? The long copulatory bouts and the sexual receptivity throughout the menstrual cycle were supposed to compensate for this loss of physically visible estrus, causing a male or males to remain interested in the female. This interest not only would result in the act of mating when the female was at the peak of her ovulatory cycle, and thus most fertile, but would also possibly ensure her protection and the protection of her offspring.

The scenario for the orang-utan's lack of an estrous cycle is based on the lifestyle of the orang-utan. Orang-utans, who live in the higher reaches of a dense tropical rain forest canopy, are typically widely separated from one another. A female orang-utan rarely mates, waiting seven or so years between offspring, and when she is ready for a mate, she actively solicits the attention of an adult male with whom she forms a consort pair for at least a few weeks. Given these conditions, it would seem to be physiologically wasteful to invest in an estrous cycle. No one would — or, more likely, could — see the sexual signal, which, given the infrequency of mating, would not be a very useful feature to have developed. Observers have remarked on orang-utans' extremely long copulatory bouts, but this aspect of their reproductive behavior has received less explanatory attention.

I was raised academically on the notion that behavior is very fluid and responsive to the winds of natural selection. But I have allowed myself to be convinced of the many ways in which behavioral traits or attributes can be obligatory. And so, given the mix of obligate and facultative traits that are supposed to make up an organism's behavioral repertoire, there seems to be some similarity between morphology and molecules in terms of degrees of variability.

The interpretation of the lack of physical, physiological, and behavioral estrus in humans as being of different origin from the apparently similar features of orang-utans is interesting in light of the notions of what is obligate and what is facultative. Neither of these notions is conceived in relation to the potential phylogenetic relationships among the organisms being discussed. This is not to say that these concepts are not framed in phylogenetic terms, that they are not cognizant of evolution. Rather, it would appear that there is still a distrust of behavior (in its broadest sense) as a factor in unraveling the evolutionary relationships among organisms.

Perhaps behavior is *not* a terribly useful phylogenetic indicator.

But one of the problems I see with sociobiological and behavioral approaches is that because the data are gleaned from observations in present-day ecological time, an appreciation of time depth, and thus of another phylogenetic level, is wanting.

For example, if you studied a suite of behaviors across a diversity of organisms, you could not, without additional information, know what the phylogenetic significance of observed similarities or dissimilarities was. Animals found to "behave" differently are interpreted easily enough as having adapted or evolved to exploit best the situations in which they find themselves. But the cases of such habitationally separated animals as humans and orang-utans, with such similar and unique reproductive behaviors, can be interpreted in more than one way.

The typical approach to the human and orang-utan cases is to ascribe the similarities to convergence — the same apparent solution to different demands of selection. However, another explanation for the degree of similarity between human and orang-utan sexual and reproductive physiology and behavior is that these two hominoids inherited these attributes from a common ancestor that was itself distinguished by these unique features. In other words, humans and orang-utans have "dragged" their lack of physical and behavioral estrus, and maybe even their propensity for long copulatory bouts, with them into their more recent ecological conditions. Thus, it would be inappropriate to try to explain the separate and independent evolution of these traits in both humans and orang-utans; this theoretical endeavor would be addressing the wrong phylogenetic level. Rather, if an explanation for the lack of estrus, for example, were to be sought, the questions should be directed at the level of the last common ancestor of humans and orang-utans.

The common ancestor of humans and orang-utans — and, ultimately, of all the members of that clade, including *Sivapithecus* — would probably have inhabited Africa around twenty million years ago. This is toward the end of the period during which the continent of Africa — actually, in terms of continental drift and plate tectonics, the African Plate — was moving northward toward the large plate composed of Europe and Asia. Previously, the African Plate had been isolated from contact with the Eurasian Plate by the ancestral waters of the Mediterranean Sea being broadly confluent on the east with the Indian Ocean and, on the west, with the Atlantic Ocean.

During its isolation, Africa was a lush continent, covered throughout with tropical evergreen rain forests. However, as the continent migrated northward, its extensive forest cover began to break up. This ecological fragmentation occurred predominantly in the east, where volcanism and plate activity were most active. Climatic shifts with increased seasonality affected North Africa first, as the warm Indian Ocean currents were being squeezed out by the northward surge of the African Plate.

Shortly after the emergence of the common ancestor of humans and orang-utans, the African Plate docked with the Eurasian Plate, effectively sealing off the Mediterranean Sea from the Indian Ocean and minimizing large-scale marine exchange with the Atlantic Ocean. There was, however, now a land bridge, a corridor through what is now the Near East, over which animal migration between Africa and Eurasia could take place. Along with many other mammals, primates — especially varieties of Old World monkeys and members of the orang-utan clade — spewed forth from the African continent, migrating into Turkey and then farther west into Greece, Italy, and Spain, as well as east, into India and Pakistan and eventually as far as China and southeast Asia. In my scenario, the last common ancestor of the African apes remained in Africa, as did the ultimate ancestor of the hominids, along with most of the Old World monkeys.

The common ancestor of humans and orang-utans, although characterized by those uniquenesses which set humans and orang-utans apart among living primates, would have been neither an orang-utan nor a hominid. It would probably not have displayed the enormous amount of sexual dimorphism that modern orang-utans develop as they mature, but size differences between males and females may have been more marked than in modern humans. The common human–orang-utan ancestor would probably not have had cheek flanges, throat sacs, overly elongate arms, or untethered hip joints. But neither would it have had the pelvic and hindlimb features characteristic of the hominids, *Australopithecus* and *Homo*.

Given the anatomical evidence, I think it is reasonable to conclude that the ancestor of humans and orang-utans was arboreal. How arboreal is hard to reconstruct with accuracy. But because Lucy's toe bones are quite curved, which is a feature supposedly correlated with arboreality, and because the orang-utan is definitely arboreal, the common ancestor of humans and orang-utans may have spent more time in the trees than do gorillas and perhaps even chimpanzees.

Certainly, the diet of orang-utans demands a more arboreal existence. And if Rich Kay of Duke University is correct about the dietary significance of having thick molar enamel — basically, for cracking tough nuts, among other hard food objects — or if, as Peter Rodman of the University of California at Davis has pointed out for orang-utans, bark was an important food source, we would expect that the presumably thick-molar-enameled human–orang-utan ancestor spent a good deal of time in the trees foraging for such edibles.

The ancestor of humans and orang-utans would have had the longest gestation period of all primates, and the females would have been distinguished further in their reproductive physiology, both hormonally and anatomically. Infants, grasping perhaps tenaciously to their mothers' long ruddy or blond hair, would have nursed on one of the broadly emplaced breasts. Perhaps there was a tendency for the mother to hold her infant in one arm rather than the other, as handedness may have been a manifestation of the animal's distinctive cerebral asymmetries. And there probably was a few years between births, to allow for the physical as well as cerebral development of the offspring.

One would predict that an adult male associated with more than one female. With the female lacking all aspects of an estrous cycle and, instead, having an active role sexually, it is likely that male–female interactions were quite different from those characteristic of gorillas and especially those of chimpanzees.

The degree of arboreality, as well as the birth spacing and the mother–infant interaction, may have been correlated with the lack of estrus in the female of the common ancestor of humans and orang-utans. Perhaps the development of long hair contributed as well to the evolutionary abandonment of this form of sexual signaling, by obscuring swollen genital regions. Whatever the reasons, it would nevertheless be the case that the lack of estrus and the potential for sexual receptivity at any time throughout the menstrual cycle would make for a very different animal indeed. Add to this picture male–female associations that would extend much longer than the few days of maximum fertility during the ovulatory cycle, with perhaps renewed consortships between the same individuals on future occasions, as well as mating behavior that has more to it than quick copulation for impregnation, and there is the basis for what we see in modern humans and orang-utans.

Perhaps the most difficult part of this extrapolation to comprehend is that these attributes would have characterized *all* members of the human–orang-utan clade, not just the two living forms and their last common ancestor. Thus, all potential orang-utan relatives as well as all possible hominids, from Lucy to other *Australopithecus*, and of course all species of the genus *Homo*, would have been similarly distinguished from all other primates. Whatever the group dynamics of early hominids, they would have been distinctly different from those of even the gorilla and the chimpanzee.

Is my theory viable? On many levels, the association of humans closely with orang-utans makes a lot of sense. It brings some order into the otherwise seemingly contradictory studies of fossil and living hominoids. In other ways, however, the idea of a close evolutionary relationship between humans and orang-utans violates some of our most basic concepts of how closeness of relatedness will be revealed in anatomy and especially in molecules, and how we will be able to perceive it. But whether or not a decade from now popular opinion will have shifted to embracing a shaggy, ruddy-haired, right-handed arborealist as an ancestor in our distant past, I hope that this possibility will continue to remind us that although there is reality to evolution, it has not yet divulged its innermost secrets.

Selected
Bibliography
Index

Selected
Bibliography
(with Some Annotation)

Andrews, P. J., and Cronin, J. E. 1982. The relationships of *Sivapithecus* and *Ramapithecus* and the evolution of the orang-utan. *Nature* 297: 541–46.

Andrews, P. J., and Tekkaya, I. 1980. A revision of the Turkish Miocene hominoid *Sivapithecus meteai*. *Palaeontology* 9: 85–95.

Baba, M. L., Weiss, M. L., Goodman, M., and Czelusniak, J. 1982. The case of tarsier hemoglobin. *Systematic Zoology* 31: 156–65.

Benveniste, R. E., and Todaro, G. J. 1976. Evolution of type C viral genes: evidence for an Asian origin of man. *Nature* 261: 101–8.

Brésard, B., and Bresson, F. 1983. Handedness in *Pongo pygmaeus* and *Pan troglodytes*. *Journal of Human Evolution* 12: 659–66.

Britten, R. J., and Kohne, D. E. 1968. Repeated sequences in DNA. *Science* 161: 529–40.

Broom, R. 1951. *Finding the Missing Link*. London: Watts & Co. (A charming history of the earlier discoveries of fossil hominids and the controversies surrounding Dart and *Australopithecus*.)

Brown, W. M., George, M., Jr., and Wilson, A. C. 1979. Rapid evolution of mitochondrial DNA. *Proceedings of the National Academy of Sciences, U.S.A.* 76: 1967–71.

Brown, W. M., Prager, E. M., Wang, A., and Wilson, A. C. 1982. Mitochondrial DNA sequences of primates: tempo and mode of evolution. *Journal of Molecular Evolution* 18: 225–39.

Bruce, E. J., and Ayala, F. J. 1979. Phylogenetic relationships between man and the apes: electrophoretic evidence. *Evolution* 33: 1040–56.

[Buffon, Comte de.] *Buffon's Natural History of Man, the Globe, and of Quadrupeds*, Volume 1. New York: Hurst & Co. (My edition has the date torn out.)

Cartmill, M. 1974. Rethinking primate origins. *Science* 184: 436–43.

———. 1982. T-lymphocyte immunology and hominoid evolution. *Science* 218: 1145.

Cave, A. J. E., and Haines, R. W. 1940. The paranasal sinuses of the anthropoid apes. *Journal of Anatomy* 74: 493–523.

Ciochon, R. L., and Corruccini, R. S., eds. 1983. *New Interpretations of Ape and Human Ancestry.* New York: Plenum Press. (Contains articles by Andrews, Pilbeam and Ward, and Mai.)

Clark, W. E. Le Gros. 1962. *The Antecedents of Man,* 2d ed. Edinburgh: Edinburgh University Press.

Cronin, J. E., and Sarich, V. M. 1980. Tupaiid and Archonta phylogeny: the macromolecular evidence. In W. P. Luckett, ed., *Comparative Biology and Evolutionary Relationships of Tree Shrews,* pp. 293–312. New York: Plenum Press.

Current Anthropology. 1965. Vol. 6, no. 4. (Contains reprints of important papers on East African paleoanthropology.)

Czekala, N. M., Benirschke, K., McClure, H. M., and Lasley, B. L. 1983. Urinary estrogen excretion during pregnancy in the gorilla (*Gorilla gorilla*), orangutan (*Pongo pygmaeus*) and the human (*Homo sapiens*). *Biology of Reproduction* 28: 289–94.

Dart, R. A. 1925. *Australopithecus africanus:* the man-ape of South Africa. *Nature* 115: 195–99.

Darwin, C. 1859. *On the Origin of Species.* (Harvard University Press facsimile of the first edition, 1964.)

———. 1871. *The Descent of Man,* Volume 1. London: John Murray.

De Beer, G. R. 1930. *Embryology and Evolution.* Oxford: Clarendon Press.

Delson, E., and Andrews, P. J. 1975. Evolution and interrelationships of the catarrhine primates. In W. P. Luckett and F. S. Szalay, eds., *Phylogeny of the Primates,* pp. 405–46. New York: Plenum Press.

Eiseley, L. 1961. *Darwin's Century.* Garden City, NY: Anchor Books. (An indispensable review of the development of evolutionary and Darwinian thought.)

Ferris, S. D., Wilson, A. C., and Brown, W. M. 1981. Evolutionary tree for apes and humans based on cleavage maps of mitochondrial DNA. *Proceedings of the National Academy of Sciences, U.S.A.* 78: 2431–36.

Galdikas, B. M. F. 1979. Orangutan adaptation at Tanjung Puting Reserve: mating and ecology. In D. A. Hamburg and E. R. McCown, eds., *The Great Apes,* pp. 195–233. Menlo Park, Calif.: Benjamin/Cummings Publishing Co. (A good review volume on great ape behavior and ecology, with papers also by MacKinnon and Rodman.)

———. 1982. Orang utans as seed dispersers at Tanjung Puting, Central

Kalimantan: implications for conservation. In L. E. M. de Boer, ed., *The Orang utan: Its Biology and Conservation*, pp. 285–98. The Hague: Dr. W. Junk Publishers. (Another solid volume on the orang-utan, with articles by Schürmann and Rijksen.)

Gaulin, S. J. C. 1979. A Jarman-Bell model of primate feeding niches. *Human Ecology* 7: 1–20.

Gaulin, S. J. C., and Sailer, L. D. 1984. Sexual dimorphism in weight among the Primates: the relative impact of allometry and sexual selection. *International Journal of Primatology* 5: 515–35.

Goodman, M. 1962. Immunochemistry of the primates and primate evolution. *Annals of the New York Academy of Sciences* 102: 219–34.

Goodman, M., and Tashian, R. E., eds. 1976. *Molecular Anthropology.* New York: Plenum Press.

Goodman, M., Braunitzer, G., Stangl, A., and Schrank, B. 1983. Evidence on human origins from haemoglobins of African apes. *Nature* 303: 546–48.

Goodman, M., Olson, C. B., Beeber, J. E., and Czelusniak, J. 1982. New perspectives in the molecular biological analysis of mammalian phylogeny. *Acta Zoologica Fennica* 169: 1–73.

Gould, S. J. 1977. *Ontogeny and Phylogeny.* Cambridge, Mass.: Belknap Press of Harvard University Press. (Along with de Beer and Haeckel, a useful reference on the development of embryological thought and the embryologists involved, especially von Baer.)

———. 1983. Chimp on the chain. *Natural History* 12/83: 18–27.

Graham, C. E., ed. 1981. *Reproductive Biology of the Great Apes.* New York: Plenum Press. (An important collection of papers by Graham, Nadler, and others on menstrual cycles, estrus, and sexual behavior.)

Gregory, W. K. 1910. The order of mammals. *Bulletin of the American Museum of Natural History* 27: 1–524. (A wonderful review of the history of classifications and their classifiers.)

———. 1915. Is *Sivapithecus* Pilgrim an ancestor of man? *Science* 42: 341–42.

———. 1922. *The Origin and Evolution of the Human Dentition.* Baltimore: Williams and Wilkins.

———. 1927. Dawn-man or ape? *Scientific American,* September.

Haeckel, E. 1896. *The Evolution of Man,* Volumes 1 and 2 (translated from the German). New York: D. Appleton and Company.

Hasegawa, M., Kishino, H., and Yano, T. 1985. Dating of the human-ape splitting by a molecular clock of mitochondrial DNA. *Journal of Molecular Evolution* 22: 160–74.

Huxley, T. H. 1863. *Man's Place in Nature.* New York: D. Appleton and Company.

———. 1896. *Man's Place in Nature.* New York: D. Appleton and Company. (Contains additional essays.)

Jones, F. W. 1929. *Man's Place Among the Mammals.* London: Edward Arnold.

King, M-C., and Wilson, A. C. 1975. Evolution at two levels in humans and chimpanzees. *Science* 188: 107–88.

Leakey, L. S. B., Prost, J., and Prost, S., eds. 1971. *Adam or Ape: a sourcebook of discoveries about early man.* Cambridge, Mass.: Schenkman Publishing Company. (An excellent collection of historically important paleoanthropological essays.)

Lowenstein, J. M., Sarich, V. M., and Richardson, B. J. 1981. Albumin systematics of the extinct mammoth and Tasmanian wolf. *Nature* 291: 409–11.

Lowenstein, J. M., Molleson, T., and Washburn, S. L. 1982. Piltdown jaw confirmed as orang. *Nature* 299: 294.

Lumsden, A. G. S. 1979. Pattern formation in the molar dentition of the mouse. *Journale Biologique Buccale* 7: 77–103.

MacKinnon, J. 1974. The behaviour and ecology of wild orang-utans (*Pongo pygmaeus*). *Animal Behavior* 22: 3–74.

Maier, W. 1981. Nasal structures in Old and New World primates. In R. L. Ciochon and A. B. Chiarelli, eds., *Evolutionary Biology of the New World Monkeys and Continental Drift,* pp. 219–41. New York: Plenum Press.

Maple, T. 1980. *Orangutan Behavior.* New York: Academic Press.

Martin, L. 1985. Significance of enamel thickness in hominoid evolution. *Nature* 314: 260–63.

Martin, R. D. 1968. Toward a new definition of Primates. *Man* 3: 377–401.

Mivart, St. G. 1864. Notes on the crania and dentition of the Lemuridae. *Proceedings of the Zoological Society of London 1864:* 611–48.

———. 1873. On *Lepilemur* and *Cheirogaleus* and on the zoological rank of the Lemuroidea. *Proceedings of the Zoological Society of London 1873:* 484–518.

Monboddo, J. B., Lord. 1774. *Of the Origin and Progress of Language,* Volume 1, 2d ed. (Reprinted New York: AMS Press, 1973.)

Nuttall, G. H. F. 1904. *Blood Immunity and Blood Relationship.* Cambridge: Cambridge University Press.

Osborn, J. 1978. Morphogenetic gradients: fields versus clones. In P. M. Butler and K. A. Joysey, eds., *Development, Function and Evolution of Teeth,* pp. 171–201. New York: Academic Press.

Peacock, T. L. 1891. *Melincourt, or Sir Oran Haut-Ton.* Edited by R. Garnett. London: J. M. Dent & Co. (First edition, 1856.)

Pilbeam, D. 1982. New hominoid skull material from the Miocene of Pakistan. *Nature* 295: 232–34.

———. 1984. The descent of hominoids and hominids. *Scientific American* 250: 84–97. (An easy-to-read summary of events surrounding the discovery of new *Sivapithecus* material.)

Romero-Herrera, A. E., Lehmann, H., Castillo, O., Joysey, K. A., and Friday, A. E. 1976. Myoglobin of the orangutan as a phylogenetic enigma. *Nature* 261: 162–64.

Romero-Herrera, A. E., Lehmann, H., Joysey, K. A., and Friday, A. E. 1978. On the evolution of myoglobin. *Philosophical Transactions of the Royal Society of London, Series B (Biological Sciences)* 283: 61–183.

Sarich, V. M. 1971. A molecular approach to the question of human origins. In P. Dolhinow and V. M. Sarich, eds., *Background for Man*, pp. 60–81. Boston: Little, Brown.

Sarich, V. M., and Wilson, A. C. 1966. Quantitative immunochemistry and the evolution of primate albumins: micro-complement fixation. *Science* 154: 1563–66.

———. 1967a. Rates of albumin evolution in primates. *Proceedings of the National Academy of Sciences, U.S.A.* 58: 142–48.

———. 1967b. Immunological time scale for hominid evolution. *Science* 158: 1200–1203.

Schultz, A. H. 1936. Characters common to higher primates and characters specific for man. *Quarterly Review of Biology* 11: 259–83, 425–55. (A superb review of the history since T. Huxley on the systematic treatment of the hominoids, including Schultz's debate with Weinert.)

———. 1963. Age changes, sex differences, and variability as factors in the classification of primates. In S. L. Washburn, ed., *Classification and Human Evolution*, pp. 85–115. Chicago: Aldine.

———. 1968. The recent hominoid primates. In S. L. Washburn and P. C. Jay, eds., *Perspectives on Human Evolution*, Volume 1, pp. 122–95. New York: Holt, Rinehart and Winston. (*The* source for a summary of Schultz's data and an exhaustive bibliography of his work.)

Schwartz, J. H. 1983. Palatine fenestrae, the orangutan, and hominoid evolution. *Primates* 24: 231–40.

———. 1984a. What is a tarsier? In N. Eldredge and S. M. Stanley, eds., *Living Fossils*, pp. 38–49. New York: Springer Verlag.

———. 1984b. The evolutionary relationships of man and orang-utans. *Nature* 308: 501–5.

———. 1984c. Hominoid evolution: a review and a reassessment. *Current Anthropology* 25: 655–72.

———. 1985. Toward a synthetic analysis of hominoid phylogeny. In P. V. Tobias, ed., *The Past, Present and Future of Hominid Evolution*, pp. 265–69. New York: Alan R. Liss.

———. 1986. Non-hominid catarrhine evolution. *South African Journal of Science* 82: 90–92.

————. In press. Primate systematics and a classification of the order. In D. R. Swindler, ed., *Comparative Primate Biology,* Volume 1: *Systematics, Evolution and Anatomy.* New York: Alan R. Liss.

Schwartz, J. H., and Tattersall, I. 1985. Evolutionary relationships of living lemurs and lorises (Mammalia, Primates) and their potential affinities with European Eocene Adapidae. *Anthropological Papers of the American Museum of Natural History* 60: 1–100.

Sibley, C. G., and Ahlquist, J. E. 1983. Phylogeny and classification of birds based on the data of DNA-DNA hybridization. In R. F. Johnston, ed., *Current Ornithology,* Volume 1, pp. 245–92. New York: Plenum Press.

————. 1984. The phylogeny of the hominoid primates, as indicated by DNA-DNA hybridization. *Journal of Molecular Evolution* 20: 2–15.

Simons, E. L. 1961. The phyletic position of *Ramapithecus. Postilla* 57: 1–9.

————. 1969. The origin and radiation of the Primates. *Annals of the New York Academy of Sciences* 167: 319–31.

————. 1972. *Primate Evolution.* New York: Macmillan Publishing Co. (At the introductory level, the best single historical review and systematic overview of nonhominoid fossil primates.)

Simons, E. L., and Pilbeam, D. 1965. Preliminary revision of the Dryopithecinae (Pongidae, Anthropoidea). *Folia Primatologica* 3: 81–152. (A detailed review of the fossil ape who's who, when the apes were still thought of as a group.)

Simpson, G. G. 1940. Studies on the earliest primates. *Bulletin of the American Museum of Natural History* 77: 185–212. (Deals with the dental issue of what a primate is.)

————. 1945. The principles of classification and a classification of the mammals. *Bulletin of the American Museum of Natural History* 85: 1–350.

Tattersall, I. 1975. *The Evolutionary Significance of Ramapithecus.* Minneapolis: Burgess Publishing Co. (A solid, introductory-level historical review, when *Ramapithecus* was still considered a hominid.)

Tuttle, R. 1974. Darwin's apes, dental apes, and the descent of man: normal science in evolutionary anthropology. *Current Anthropology* 15: 389–98. (An excellent review of the theories of brachiation and knuckle walking preceding human bipedalism as well as the morphologies of knuckle walking.)

Tyson, E. 1699. *Orang-outang, sive Homo sylvestris; or, The Anatomy of a Pygmie Compared with That of a Monkey, an Ape, and a Man.* London.

Vrba, E. S. 1979. A new study of the scapula of *Australopithecus africanus* from Sterkfontein. *American Journal of Physical Anthropology* 51: 117–30.

Wallace, A. R. 1898. *The Malay Archipelago: The Land of the Orang-utan and the Bird of Paradise.* London: MacMillan and Co.

Ward, S. C. 1983. Canine implantation in Miocene hominoids. *American Journal of Physical Anthropology* 60: 268.

Ward, S. C., and Kimbel, W. H. 1983. Subnasal alveolar morphology and the systematic position of *Sivapithecus*. *American Journal of Physical Anthropology* 61: 157–72.

White, T. D., Johanson, D. C., and Kimbel, W. H. 1981. *Australopithecus africanus:* its phyletic position reconsidered. *South African Journal of Science* 77: 445–70.

Yunis, J. J., and Prakash, O. 1982. The origin of man: a chromosomal pictorial legacy. *Science* 215: 1525–30.

Zuckerkandl, E., and Pauling, L. 1962. Molecular disease, evolution, and genic heterogeneity. In M. Kasha and B. Pullman, eds., *Horizons in Biochemistry*, pp. 189–225. New York: Academic Press.

Index

acclimatization: behavior and, 300
Adapis, 26, 27
adrenal glands, 164–65
Afar Triangle: evolutionary research in, 53–54
Africa: ecological fragmentation of, 303–4; evolutionary research in, 24–25, 30, 42–56
African apes: "African pattern" nasal cavity in, 190–94; brow ridge development in, 195; canines of, 200–202; carpal bones in, 188–89; chromosome banding patterns in, 274–76; divergence from hominids, dating, 234–35; DNA hybridization research on, 262, 264; evolution from *Proconsul*, 78–79; fibrinopeptide protein sequence in, 292; hand muscles of, 92–93; and hominid evolution, 85; and hominoid relatedness, 57, 64–69, 71, 113–20, 232; ischial callosities in, 210; knuckle walking by, 7–8, 91–94; legends about, 18; molar structure of, 203–4, 285–88; molecular research on, 81, 226, 227, 228–29, 234, 266, 272, 273–74; morphological affinities to hominids, 181–218;

myoglobin research on, 249–51, 266; names for, 17–18; 19th century evolutionary research on, 23–25, 30; nucleotide mismatch with *Homo*, 261–62; palatine ridges in, 187; pattern 1 enamel in, 284–88; sinuses in, 182–84; talus in, 199; and 20th century evolutionary theory, 56; *see also* chimpanzees; gorillas; hominoid(s); hominoid evolution; hominoid relatedness
"African pattern" nasal cavity, 190–94
African Plate, 303–4
Ahlquist, Jon, 262–64, 265, 271, 272, 295
AIDS, 221
albumin: and hominoid evolution/relatedness, 228, 229, 230, 231, 233, 239; of mammoth, 277–78; of Tasmanian wolf, 278
American Museum of Natural History: "Ancestors" exhibit, 217
amino acids: and fibrinopeptide protein sequence, 292; *see also* proteins
Anaptomorphus, 90
Andrews, Peter, 83, 177, 178–79, 181–82, 185, 202–3, 217, 261, 292, 293

androgens, 164, 215–16

androsterone, 164, 215–16

Angola: exploration of, 17

anthropoids (Anthropoidea): dominance of sight over smell in, 145–46, 149; frontal suture in, 150; hands and feet of, 141; hemochorial placentation of, 150–51; mandibular structure of, 150; molecular approach to relationships among, 226, 227; myoglobin research on, 249; postorbital bar in, 133, 141; and prosimians, evolutionary relationship, 147; shared characteristics of, 151; simplex uterus in, 150, 151; as suborder, 127; *see also* chimpanzees; gibbons; gorillas; hominoid(s); *Homo;* New World monkeys; Old World monkeys; orang-utans; primate(s)

Anthropomorpha, 122

antibodies, 222–23

antiserum, 223–26, 227; serum-antiserum reaction, 223–26, 227, 229, 230, 279–81

antitoxins, 223

apes. *See* African apes; chimpanzees; gibbons; gorillas; hominoid(s); orang-utans

Apidium, 143

Arambourg, C., 48

arborealism: of common orang-utan/hominid ancestor, 304–5; and scapular structure, 198

Archencephala, 131

Aristotle, 15–16

arms: bone ossification in, 161–62 (*see also* ossification); of hominoids, 154–55; *see also* scapulae; wrist structure

"Asian pattern" nasal cavity, 190–94

Assyria, ancient: taxonomy in, 98

Australopithecinae, 102

Australopithecus: and "African pattern" nasal cavity, 190, 191–94; canines of, 97, 200–201; dating, 296; and hominid evolution, 79–80, 81, 84–85, 87, 139, 175, 232, 235–36, 296, 304, 306; scapulae of, 198; single incisive foramen in, 177, 191–93; taxonomic classification of, 101–2, 106; third molar in, 202–3

A. afarensis, 54–56, 106, 177, 191, 205, 296; basicranial region of, 205–7; dating, 296; *see also* Lucy

A. africanus, 44–45, 47–49, 101, 106, 296

A. boisei, 50–52, 106, 296

A. robustus, 48–49, 296

Ayala, Francisco, 254–55, 273, 294

aye-aye (*Daubentonia*), 103–5, 112–13, 117, 126; claws of, 132; taxonomic classification of, 103–5, 117, 127; teeth of, 148; *see also* lemurs

Aziz, M. Ashraf, 289–90

baboons, 91, 115; arm bone ossification in, 161; bilophodonty in, 128; canines of, 95; DNA of, 261; fossils, 143; ischial callosities in, 209–10; myoglobin research on, 249; taxonomic classification of, 103; tooth structure of, 95, 128, 166; *see also* Old World monkeys

basicranial region: hominoid, 205–7; *see also* skull structure

bats, 100; cerebrum of, 130; Linnaean classification of, 122, 123; mammary glands of, 131; tooth structure of, 131

beaver, 104

behavior: facultative traits, 300–301, 302; morphological outlook on, 300–303; obligate traits, 300–301, 302; phylogenetic sig-

nificance of, 300–303; in taxonomic classification, 100

Benveniste, Raoul, 260–61

bilophodonty, 128, 159

Bimanes (Bimana), 123

biochemical approach to evolution,
226–82; *see also* molecular research

Biogenetic Law, 61–62

bipedalism: and brachiation, 158;
and evolutionary theory, 36–37,
47, 48, 54–55, 88, 90–91; and
knuckle walking, 91, 95

birds: DNA hybridization and phylogeny of, 262–64, 272

Black, Davidson, 41, 42

Blainville, Henri Marie Ducrotay
de, 28, 130

blood groups, 299

*Blood Immunity and Blood Rela-
tionships* (Nuttall), 221

blood proteins. *See* proteins

blood Rh factor, 221–23

blood serum research, 253–55; albumin, 228, 229, 230, 231, 233,
239, 277–78; ceruloplasmin,
228; electrophoresis, 228;
gamma globulin, 228, 229;
hemoglobin, 239–42, 244–48,
264, 266, 299; and hominoid relatedness, 228–29; macroglobulin, 228; myoglobin, 247–51,
266; serum-antiserum reactions,
223–26, 227, 229, 230–31,
279–81; transferrin, 228, 229,
239

Blumenbach, Johann Friederick,
123

Bonaparte, Prince Charles Lucien,
130, 131

Bontius, 16–17

borhyaenids, 278

Borneo, 16

Bornean orang-utans, 11, 16; description of, 11; face of, 8–9; *see
also* orang-utans

Bory St. Vincent, Jean Baptiste
Georges, 20–21

Boule, Marcelin, 36

brachiation, 156–58; and bipedalism, 158

brachiators, 88, 90

Bradypus. See sloths

brain asymmetries, 165, 213–16,
284, 300, 305

brain development: asymmetries,
see brain asymmetries; and
classification of primates, 130–
31; and evolutionary theory, 52,
59–60, 66, 95–97, 165; and
tool use, 95–96; *see also* skull
structure

Bramapithecus, 75, 77

B. punjabicus, 75, 77

B. thorpei, 75

breasts: pendent, in hominoids,
185; *see also* mammary glands

Brésard, B., 214–15

Bresson, F., 214–15

Britten, R. J., 258–59, 291

Broom, Robert, 45, 70

Brown, Barnum, 75

Brown, Wes, 268–71, 272, 273

brow ridge development: and hominoid relatedness, 195–96

Bruce, Elizabeth, 254–55, 273, 294

Buffon, Georges Louis Leclerc de,
27

Buffon's Natural History, 9–10

Burnett, James (Lord Monboddo),
16–17, 21

bushbabies: clawlike "nail" or
grooming claw of, 147; Flower's
classification of, 127; mating
habits of, 13; molecular vs. morphologic research on, 237–38;
myoglobin research on, 249;
tooth comb of, 148

caecum, 131

Callithricidae, 128

camel: anatomy of, 98, 99

canines: in Catarrhini, 153; in definition of primates, 123, 129–30;
and evolutionary theory, 40, 73,
74, 95–97; hominid, 200–201;
hominoid, 166–67, 200–201; of
mammals, 131; Piltdown, 279–
81; rotation of, 200–202; *see
also* tooth structure
capuchin monkeys: molecular research on, 230
carnivores: cerebrum of, 130;
myoglobin research on, 248, 251
carpal bone: hominoid, 155, 159,
162, 163, 173, 188–89
Carthage, 99
Cartmill, Matt, 135–37, 146, 218,
273–74
catarrhines (Catarrhini), 152–53,
156, 159–60, 161, 163, 165,
166–67; DNA hybridization in,
260; molecular research on, 226
catarrhinism, 152–53
catastrophism, 26–27
cats, 132–33
caudal vertebrae: hominoid, 160
Cave, A. J. E., 184, 195
Cebidae, 128
cercopithecids (Cercopithecidae).
See Old World monkeys
cerebral asymmetry, 165, 213–16,
284, 300, 305; *see also* brain development
ceruloplasmin, 228
cheek teeth. *See* tooth structure
chickens: myoglobin research on,
251
chimpanzees (*Pan*): African legends
about, 18; African names for,
18; and "African pattern" nasal
cavity, 190; antiserum testing
with, 279–81; basicranial region
of, 205–6; brain of, 66; brain
asymmetry in, 214; canines of,
166–67, 200–201; chromosome
bonding pattern in, 274–76; and
classification of orang-utan, 16;

closeness to *Homo*, molecular research on, 85, 246–47, 249–51,
254, 264, 266, 273; dating taxonomic divergence from *Homo*,
296; DNA of, 261; DNA hybridization research on, 264; early
reports of, 17; fibrinopeptide
protein sequence in, 292; genital
swelling in estrus, 211; genomic
similarity to *Homo*, 295–96;
gestation period of, 210–11;
Goodall's research on, 1, 2; and
great chain of being, 19–20; hair
density in, 194–95; handedness
in, 215; hands and feet of, 7–8,
92–94; hemoglobin of, 246–47,
266; and hominoid relatedness,
65–69, 71, 85, 113, 294–96; ischial callosities in, 210; knuckle
walking by, 92–95; mating habits of, 13–14, 15; molar class
stem cells of, 299; molar structure of, 203–4; molecular research on, 81, 85, 226, 227,
228–29, 230, 246–51, 253–54,
255, 264, 266, 272, 273; myoglobin research on, 249–51; and
19th century evolutionary research, 23–25; palatine fenestrae
in, 168–69, 191; pattern 1
enamel in, 285–88; polygyny
among, 301; and *Proconsul*, 78;
regulatory genes in, 254; relatedness to gorilla, 68, 113; sinuses
in, 183–84; toes of, 65; trunk
structure of, 153; and 20th century evolutionary research, 56;
see also African apes; homi-
noid(s); hominoid evolution;
hominoid relatedness; primate(s);
pygmy chimpanzees
Chimpenza, 18
Chiromyidae, 127
Chiromyoides, 105
Chiromys, 105, 126; *see also* aye-
aye

Chou Kou Tien cave site, 41, 42
chromosome banding patterns:
  hominoid, 274–76
cingula: molar, 203–4
cladism, 111
Clark, J. Desmond, 276–77
Clarke, Dr., 10
classification. *See* taxonomy
clavicle, 131–32; hominoid, 123,
  129, 131–32, 140, 155, 159
claws: vs. nails, in primates, 130
cloven-footed animals, 98, 99
coccygeal vertebrae: hominoid,
  160
codons, 240–41
Colobus, 128
copulating behavior: of common
  orang-utan/hominid ancestor,
  305; duration of copulatory bout
  in hominoids, 301–2; obligate,
  301; among primates, 13–15;
  and sexual dimorphism, 157;
  similarities between orang-utan
  and *Homo*, 300–303; *see also*
  estriol secretion; estrous cycle
cows, 99
cranial structure. *See* skull structure
Crick, Francis H. C., 226
Cronin, Jack, 182, 195, 232, 237,
  239–40, 246, 261, 292
crows: tool use by, 96
cud chewing, 99
Cuvier, Georges, 26–27, 123, 125,
  129–30
cysteine, 249–50
Czekala, Nancy, 165, 215–16

Dart, Raymond, 42–45, 101, 102,
  139
Darwin, Charles, 3, 22, 27, 30, 36,
  44, 45, 58, 72, 85, 87, 91, 225;
  Descent of Man, 22–24, 30, 58,
  71; *On the Origin of Species*,
  22, 27, 30, 57, 59
dasyurids, 278
Daubentonia. *See* aye-aye

Daubentonioidea, 105
Dawson, Charles, 37, 39, 40
deciduous teeth, 204–5
Delson, Eric, 190, 202–3
de Maillet, Benoit, 6, 17
dental structure. *See* tooth structure
deoxyhydronucleic acid. *see* DNA
Descent of Man (Darwin), 22–24,
  30, 58, 71
diet: animal classification by, 100
digits: big toe, 132, 133, 141; of
  mammals, 132; of primates, 130,
  132; thumbs, 65, 132, 141
DNA (deoxyhydronucleic acid),
  226, 240, 256–65; mitochondrial, 268–72, 299; and morphologic inconsistencies, 297;
  nuclear, 268–69; and proteins,
  256; sequences, 256–61, 291,
  297
DNA hybridization, 253, 256–65,
  272, 291–93, 295
dolphins: myoglobin research on,
  248
Dryopithecus: and hominid evolution via *Ramapithecus*, 79;
  and hominoid evolution, 29–33,
  55, 72–74, 77–79, 82, 143,
  232; taxonomic classification
  of, 102
D. fontani, 29–33, 55, 78
D. giganteus, 73
D. punjabicus, 73, 75, 77
Dubois, Eugène, 33–35, 41, 44, 69,
  139
Dunlap, Samuel, 289–90
Dutch East India Company, 6
Dyaks (Bornean tribe), 11

ear lobes: in hominoids, 187
ear structure: petrosal bone, 206–7
East Turkana: evolutionary research at, 53
Educabilia, 130, 131
elbow joint: hominoid, 155, 159,
  161

Eldredge, Niles, 208, 294
electrophoresis, 228
elephant: relationship to mammoth, 277–78
elephant shrews, 137
Elliott, D. G., 10
embryology: and evolutionary research, 60–63; "stem" cells, 298; *see also* fetal development
enamel: molar. *See* molar structure
Enjocko, 18
Eoanthropus dawsoni (Piltdown Man), 36–40, 41, 70, 278–81
epiphyses, 160–64, 198, 284, 299–300
Erecta, 125
estriol secretion: hominoid, 164–65, 173, 215–16, 284, 300
estrogens, 164, 170
estrous cycle, 13, 211–12, 301–2; and genital swelling, 211–12, 301–2, 305
Ethiopia: evolutionary research in, 53–54
ethmoid bone, 183–84
Eurasian Plate, 303–4
evolutionary research and theory, 26–56, 60–63, 70–97, 108–12, 137–55, 219–82; *see also* hominid evolution; hominoid evolution; primate evolution; *and specific researchers*
extensor muscles and tendons of hand, 92
eye: postorbital bars, 132–33, 141, 145–46, 147, 149, 151; sight in evolution of primates, 145–46, 149

facultative traits, 300–301, 302
family: in taxonomy, 101
Fayum, Egypt, fossil deposits, 142–43
femur: and evolutionary theory, 35, 36; hominoid, 160–61
fetal development: and estriol

levels, 165, 173; and evolutionary research, 60–63; and gestation period, 210–11, 233; hominoid, 215–16; and immune system, 221–23; Rh factor, 221–23
fibrinopeptides, 247; protein sequences, 292
fish: hemoglobin sequence in, 244
Fleagle, John, 290
flexor muscle of hand, 92–94
Flower, Sir William Henry, 59–60, 125–29
flycatchers, 263
flying lemurs, 137; mammary glands of, 131; molecular research on, 240
Fontan, M., 29
foot: cloven, 98, 99; fossil research, 50–52; grasping, in primates, 132, 141, 147; hominoid, 199
foramen lacerum, 206–7, 208
Ford, Susan, 128
Fossey, Dian, 2
fossil: making of a, 108
fossil research, 26–56, 70–85, 90, 106–12; *see also* evolutionary research and theory; *and specific hominid and hominoid fossils*
Fraipont, J., 32–33
Friday, A. E., 265
frontal sinuses: hominoid, 71–72, 182–84, 186–87, 195
frontal suture, 150

Galdikas, Biruté, 2, 11–12
gamma globulin, 228, 229
genes: regulatory, 253–54, 293–94; structural, 253; type C viral, 26
genetic distances: among hominoids, 254–56, 293–94, 299
genetic reservoir: transmission of, 219–20
genetics: and hominoid relatedness, 226–52; *see also* molecular research
genetic transference, 219–20

genital swelling: in hominoids,
211–12, 301–2, 305
genomes, 219–20, 258–59, 291,
293–94, 295; and morphology,
294–96; uniform average rate
(UAR) of change, 262–63
genus: in taxonomy, 101
Gervais, François Louis Paul, 28
Geschwind, Norman, 165, 214–15
gestation period: of common
orang-utan/hominid ancestor,
305; in hominoids, 210–11, 233,
284, 300; and molar enamel, 286
gibbons (Hylobatidae): anatomical
characteristics of, 157; as brachiators, 156–58; brow ridge in,
195–96; canines of, 166–67;
carpal bones of, 189; chromosomes of, 275–76; distribution
of, 157; DNA hybridization research on, 264; habitat of, 157;
and hominoid relatedness, 67,
68, 69, 78, 115, 158–59; and
human bipedalism, 88, 90–91;
ischial callosities in, 210; mating
habits of, 157, 301; molar structure of, 166, 202, 203, 284–88;
molecular research on, 228–29,
230, 231, 234, 237, 247–48,
254, 274; monogamy among,
301; myoglobin research on,
249–51; nasal structure of, 167;
ossification in, 163, 198; palatine
fenestrae of, 168–69, 171; palatine ridges in, 187; and *Pliopithecus antiquus*, 28; sacral
segments in, 159; serum of, 230;
trunk structure of, 153; type C
viral gene in, 260
Gigantopithecus, 78, 79, 80, 81,
143, 175, 203–4; canines of, 97;
molars of, 284; and *Sivapithecus*,
176; taxonomic classification of,
102; tooth structure of, 97, 175,
177, 284
Gingerich, Philip, 143

glycine, 250–51
Goodall, Jane, 1, 2, 57
Goodman, Morris, 82, 226–29,
231–33, 235, 237–40, 245–47,
249, 254, 266
gorillas (*Gorilla*): basicranial region
of, 206; and brachiation, 156;
brain of, 66; brain asymmetry in,
214; canines of, 166–67, 200–
201; chromosome banding pattern in, 274–76; and classification of orang-utan, 16; DNA of,
261; DNA hybridization research
on, 264; early reports of, 17;
fibrinopeptide protein sequence
in, 292; genital swelling in, 211;
and genomic classification, 295–
96; gestation period of, 210–11;
and great chain of being, 19;
hand muscles of, 93–94; hands
and feet of, 7–8; hemoglobin in,
244, 248; and hominoid relatedness, 64–70, 71, 113; ischial
callosities in, 210; known as Impungu, 17–18; knuckle walking
by, 92, 94–95; mating habits of,
13–15, 301; molar class stem
cells in, 299; molar structure of,
203–4; molecular research on,
81, 85, 226, 227, 228–29, 255,
272, 274–76; mountain, 2; nasal
structure of, 167–68; 19th century evolutionary research on,
23–25; palatine fenestrae of,
168–69, 191; polygyny among,
301; and *Proconsul*, 78; pattern 1
enamel in, 285–88; relatedness
to chimpanzee, 68, 113; sexual
dimorphism among, 19; sinuses
of, 182–84; skull formation,
male vs. female, 19; thumb of,
65; toe of, 65; tooth structure
of, 76, 166–67, 200–204, 285–
88, 299; trunk structure of, 153;
and 20th century evolutionary
theory, 56; *see also* African apes;

gorillas (*cont.*)
  hominoid(s); hominoid evolution; hominoid relatedness;
  primate(s)
Gould, Stephen Jay, 19–20, 60, 87,
  294
Graham, Charles, 211–12
great apes: evolved from *Dryopithecus,* 78–79; field studies on,
  origins of, 1–2; relatedness, 68–
  69; tool use by, 96; *see also* African apes; chimpanzees; gorillas;
  hominoid(s); orang-utans;
  primate(s)
great chain of being, 16, 19, 72,
  127, 185
Gregory, William King, 68–69,
  70–72, 74, 75, 85, 90, 113, 122,
  184–85, 229, 290
grooming claw, 147–49; *see also*
  bushbabies; lemurs; lorises;
  mouse lemurs; tarsiers
Groves, Colin, 10
Gyrencephala, 131

habitus: animal classification by, 100
Hadar: evolutionary research at,
  53–54
Haeckel, Ernst, 61–62, 130, 137
hair density: and hominoid relatedness, 194–95
hair length: in orang-utan and
  Homo, 115, 196, 284, 305
hand(s): grasping, 132, 141; in
  hominid classification, 123, 125;
  of primates, 122, 123, 130, 132,
  141, 147
handedness: and brain asymmetry,
  213–15, 305; in hominoids,
  214–15
hand muscles: of African apes, 92–
  94
Hapalidae, 128
hedgehogs, 137, 139; myoglobin
  research on, 251
Hellman, Milo, 75

hemochorial placentation, 150–51,
  233
hemoglobin, 239–42, 244–48, 264,
  266; chains/sequences, 239–40,
  244–48; variations in, 299
heredity. *See* genetics
heterochrony, 162
hierarchical ordering of organisms,
  15–16
hip joints: of primates, 6
hominid(s) (Hominidae): arbitrariness of classification, 103; brow
  ridge development in, 195–96;
  canines of, 200–202; dating divergence from other hominoids,
  234–35; definition of, 47; evolutionary research on, *see* hominid
  evolution; as family, 101–3,
  129; single incisive foramen in,
  174, 177, 178, 179, 189–91;
  tooth structure of, 175–77, 178;
  see also *Homo*
Hominidae. *See* hominid(s)
hominid evolution, 47–56; *Australopithecus,* 79–80, 81, 84–85,
  87, 139, 175, 232, 235–36, 296,
  304, 306 (see also *Australopithecus*); bipedalism, 91, 95;
  Bramapithecus, 75, 77; general
  outline of, 78–79; *Gigantopithecus,* 79–81, 143, 175, 203–4
  (see also *Gigantopithecus*);
  Kenyapithecus, 77; molecular
  research on, 81–83 (*see also*
  molecular research); *Ramapithecus,* 73–83, 174–77, 232, 235–
  36, 247 (see also *Ramapithecus*);
  Sivapithecus, 73, 74, 78, 79–81,
  83–85 (see also *Sivapithecus*);
  Tarsius and, 88–89; and taxonomic classification, 102, 106;
  tool use, 96; tooth structure, 76,
  95–97; *see also* hominoid evolution
hominid relatedness. *See* hominoid
  relatedness

Homininae, 102; see also *Homo*

hominoid(s) (Hominoidea): anthropoid characteristics of, 151; arms of, 154–55; basicranial region in, 205–7; as brachiators, 156–58; brain asymmetry in, 165, 213–16, 284, 300, 305; canines of, 166–67, 200–201; carpal bones in, 155, 159, 162, 163, 173, 188–89; catarrhinism in, 152–53, 159; chromosome banding patterns in, 274–76; clavicle of, 123, 129, 131–32, 140, 155, 159; divergence among, dating, 234–35; DNA hybridization among, 260, 272; ear lobes in, 187; elbow joint of, 155, 159, 161; estriol secretion in, 164–65, 173, 215–16, 284, 300; estrous cycle in, 13, 211–12, 301–2, 305; features distinguishing orang-utan among, 292–93; femur of, 160–61; fetal development among, 215–16; foot of, 199; fossils of, 143 (*see also* hominoid evolution); frontal sinuses in, 71–72, 182–84, 186–87, 195; genetic distances among, 254–56, 293–94, 299; genital swelling in, 211–12, 301–2, 305; gestation period among, 210–11, 233, 284, 300; gibbon and siamang as, 158–59; handedness in, 214–15; humerus of, 154, 155, 159, 161; ilial blades of, 154, 159, 165; ischial callosities in, 209–10, 284; mammary glands of, 210, 284, 305; mating behavior among, 211–12, 215–16; maxilla and premaxilla of, 168, 190–91; maxillary sinuses in, 182–83; molar structure of, 118–19, 165, 175, 178, 202–3, 283–89; myoglobin research on, 249; nasal structure of, 152–53, 159, 167, 182–85, 186, 190–94; neoteny among, 162–63; ossification in, 161–64, 173, 198, 284, 299–300; palatine fenestrae in, 168–69, 171, 173–74, 177, 178, 189–93, 205; pelvic structure in, 153–54, 159, 165; petrosal bone of, 206–7; pregnancy among, 210–11, 233, 284, 300; reproductive physiology among, 215–16; rib structure of, 154, 159; sacrum of, 159–60; scapulae of, 90, 154–55, 159, 165, 196–98, 284; sex chromosomes of, 275; sex hormones of, 164–65; shoulder structure of, 90, 154–55, 165, 196–98, 284; talus of, 199; thoracic structure of, 154; tool use by, 96; tooth structure of, 165–67, 200–203, 283–89; trunk structure of, 153–54, 156; ulna of, 155, 159, 173; vertebral structure of, 159–60; wrist structure of, 155, 159, 173, 188–89; *see also* chimpanzees; gibbons; gorillas; hominid(s); hominoid evolution; hominoid relatedness; *Homo;* orang-utans

hominoid evolution, 159–80; *Australopithecus*, see *Australopithecus; Bramapithecus*, 75; carpal bones, 188–89; distinguishing features of orang-utan, 292–93; *Dryopithecus*, 29–33, 55, 72–73, 74, 77–79, 82, 143, 232; and hemoglobin research, 239–42, 244–48; and migration from African Plate, 304; mitochondrial DNA research, 269–72; and molar structure, 284–89; molecular research on, 81–83, 226–82; morphologic and molecular inconsistencies in approach to, 297; morphologic characteristics of common orang-

hominoid evolution (*cont.*)
utan/hominid ancestor, 283–84,
304–6; *Pliopithecus*, 28–29; *Ramapithecus*, 73–83 (see also *Ramapithecus*); and relatedness, 70,
72–85, 112–20 (*see also* hominoid relatedness); *Sivapithecus*,
73, 74, 78, 79–81, 83–85 (see
also *Sivapithecus*)
hominoid relatedness, 57–85, 98–
120, 156–80; African apes and,
113–20; and antiserum, 225–
26; and brow ridge development,
195–96; current controversy
over, 283–84, 289–90; DNA
hybridization research, 262–64;
and evolution, 70, 72–85, 112–
20 (*see also* hominoid evolution); and fossil research, 108;
genetic research, 253–56; and
genomic dissimilarity, 295–96;
and hair density, 194–95; and
limb musculature, 290–91; mitochondrial DNA research on, 269–
72; and molar structure, 299;
molecular research on, 219–82;
morphologic outlook on, 181;
myoglobin research on, 249–51;
orang-utan and, 114–20, 178–
79, 283–306 (*see also* orang-
utans); sinuses, 182–84, 195
Homo: chromosome banding pattern of, 274–76; deciduous teeth
in, 205; estriol secretion and,
215–16; and estrous cycle,
211–12; fibrinopeptide protein
sequence in, 292; foramen
lacerum in, 207; frontal sinuses
in, 183–84; and genomic dissimilarity, 291–96; gestation period
in, 210–11; hair density of, 194–
95; hair length of, 196; hemoglobin of, 246–47; in Linnaean
classification, 122; mitochondrial
DNA of, 269; molar cingula in,
203–4; molar structure of, 202–

4, 285–88; molecular research
on hominoid relatedness, 226–
82; myoglobin research on, 248–
51, 266; palatine ridges in, 186–
87; polygyny among, 301; regulatory genes in, 253–54; relatedness to orang-utan, 56, 57, 67,
84–85, 113–15, 178–79, 181–
218, 283–306; scapula of, 197–
98; sexual behavior among, 212,
301–3; talus in, 199; taxonomic
classification of, 101–3, 106;
third molar in, 202–3; two-
handedness of, 123, 125; type C
viral gene in, 260; *see also* homi-
nid(s)
H. erectus, 35, 41, 86; single incisive foramen in, 177; see also
Pithecanthropus erectus
H. habilis, 52–53, 86
H. sapiens: in Linnaean classification, 20; single incisive foramen
in, 174, 177, 178, 179; see also
Homo; hominid(s)
H. sylvestris, 19–20
homology: determination of, 118–19
Homo-simiadae, 44, 101
Hopwood, Arthur Tindell, 28–29,
75
horses: eye socket structure in, 132;
hemoglobin sequence in, 244;
myoglobin research on, 248, 251
Howell, F. Clark, 50, 53
humans. *See* hominid(s); homi-
noid(s); *Homo*
humerus: epiphyseal ossification,
161, 163, 164, 198; hominoid,
154, 155, 159, 161
Huxley, Thomas, 23, 31–33, 57–
60, 63–67, 68, 70, 71, 113, 125,
130, 131, 137, 158, 225, 226,
276; *Man's Place in Nature*, 57–
60, 158, 225, 290; "On the Relatedness of Man to the Lower
Animals," 58, 125
Hylobatidae. *See* gibbons; siamangs

ilial blade: in hominoids, 154, 159, 165

Illiger, Carl, 125, 130

immune system, 221–23

immunological distance (I.D.), 230–31, 234

immunology: and evolutionary research, 55, 70–71

Impungu, 17–18

incisive foramen, single, 174, 177, 178, 179, 189–91, 284, 293; nerves in, of orang-utan, 293

incisors: in definition of primates, 129–30; of mammals, 131

index of dissimilarity, 230

Ineducabilia, 130, 131

infraorbital canal, 146

infraorder: in taxonomy, 101

insectivores (Insectivora): cerebrum of, 130; defining, 137; and evolution of primates, 137, 138–39; members of, 137–38

ischial callosities, 209–10, 284

Itsena, 18

Java, 16; evolutionary research in, 33–37

Java Man (*Pithecanthropus erectus*), 35–38, 39, 41, 44, 69

jaw structure: hominoid, 168; *see also* mandible structure; skull structure; tooth structure

Jocko, 18

Johanson, Donald C., 54, 97, 106, 135, 177, 205–7

Jolly, Clifford, 54

Jones, F. Wood, 134

Joysey, Kenneth, 248

Kalb, John, 53–54

kangaroo: myoglobin research on, 248

kangaroo rat, 88

Kay, Richard, 84, 305

Keith, Sir Arthur, 52, 131

Kenyapithecus wickeri, 77

Kimbell, William, 190, 201, 205–7

King, Mary-Claire, 253–54, 291, 294

Kingsley, Susan, 210

knuckle walking, 7–8, 178; anatomy of, 94–95; and bipedalism, 91; and evolutionary theory, 91–94

Koenigswald, G. H. R. von, 78

Kohne, D. E., 258–59, 291

Kosher, Laws of, 98

Kromdraai site, 46–47, 49

La Chapelle-aux-Saints grotto, 36

Lamarck, Jean Baptiste, 27

Lartet, Edouard, 27, 29

lamprey: myoglobin research on, 248

language: brain asymmetry and development of, 213–14; 18th century research on origins of, 17, 21, 22

Lartet, Edouard, 27, 29

Lasley, Bill, 165, 215–16

Leach, Sir Edmund, 98

Leakey, Louis S. B., 48, 49–51, 76–77, 97, 106, 135

Leakey, Mary, 48, 49, 52

Leakey, Richard, 48, 50, 52–53, 106

Le Gros Clark, Sir Wilfred, 134–35, 137, 138–40

LeMay, Marjorie, 165, 214–15

lemur(s), 6, 88, 122–23; *Adapis* fossil, 26; clawlike "nail" or grooming claw of, 147; Flower's classification of, 127; fossils of, 26, 142, 143; in Linnaean classification, 122–23; as lower primate prototype, 127, 128; molecular vs. morphologic research on, 237–38; palatine ridges in, 187; postorbital bars in, 133; tooth comb of, 148

Lemuridae: members of, 127

Lemuroidea: members of, 127; Mivart on, 129

Levine, L., 230
Leviticus, 98, 99, 100
Lewin, Roger, 265
Lewis, G. Edward, 73, 75, 77
limb bud differentiation, 299
Limnopithecus, 29, 78, 143
Linnaeus, Carl von Linné, *known
   as*, 20, 122; classification of primates, 122–23, 125–26, 129–
   30, 141; *Systema Naturae*, 122,
   125, 127, 129
Lissencephala, 131
locomotion: arborealism, 198,
   304–5; bipedalism, 36–37, 47,
   48, 54–55, 88, 90–91, 95, 158;
   brachiation, 156–58; knuckle
   walking, 7–8, 91–95, 178; and
   tool use, 96
Lohest, M., 32–33
lorises, 6, 88; clawlike "nail" or
   grooming claw of, 147; Flower's
   classification of, 127; fossils of,
   142; molecular vs. morphologic
   research on, 237–38; postorbital
   bars of, 133; slow, 207–8, 251;
   tooth comb of, 148
Lorisidae: members of, 127, 207–8
Lowenstein, Jerold, 277–81
Lucy, 54–56, 97, 106, 177, 188,
   304, 306
lumbar vertebrae: hominoid, 159–
   60
Lumsden, Andrew, 298
Lydekker, Richard, 73–74
Lyell, Charles, 58

macaques, 103; myoglobin research
   on, 249; ossification in, 161
MacKinnon, John, 3, 5, 8–9, 12
macroglobulin, 228
macromolecules, 240–41, 254
Macrotarsi, 125
Madagascar: primates of, 238
Mai, Larry, 276
Malay Archipelago, The (Wallace),
   10

Malay language: etymology of
   orang-utan, 16
Malaysia, 16
mammals (Mammalia): cerebrum
   of, 130; classification of, 130–
   34; digits of, 132; palatine ridges
   in, 186–87; penis of, 132; postorbital bars in, 133; tooth structure of, 131
mammary glands, 131; albumin,
   277–78; in hominoids, 210, 284,
   305; of primates, 122, 284
mammoth: relationship to elephants, 277–78
manatees: cerebrum of, 130
mandible structure: and evolutionary theory, 28–31, 37–39, 43–
   44, 51, 54–56, 150; *see also*
   tooth structure
mandrills, 103, 211
Man's Place in Nature (Huxley),
   57–60, 158, 225, 290
marine worm: myoglobin research
   on, 248
marmosets: classification of, 128;
   myoglobin research on, 249
marsupials: borhyaenids, 278;
   dasyurids, 278; grasping foot in,
   132, 141; myoglobin research
   on, 248, 251; tooth structure of,
   131
Martin, Lawrence, 284–89
Martin, Robert D., 140, 210
Marx Brothers, 267–68
mating behavior: of common
   orang-utan/hominid ancestor,
   305; hominoid, 211–12, 215–
   16; obligate, 301; of primates,
   13–15; and sexual dimorphism,
   52–53, 157, 166, 300, 301, 304;
   similarities between orang-utan
   and *Homo*, 300–303
maxilla: hominoid, 168, 190–91,
   193
maxillary sinuses: hominoid, 182–
   83

McHenry, Henry, 199

McKitrick, Mary, 263

Melincourt; or, Sir Oran Haut-ton (Peacock), 21–22

menstrual cycle, 13, 212; and estriol secretion, 215–16; and genital swelling, 211–12; *see also* estrous cycle

messenger RNA, 226, 257

metacarpals, 99; epiphyseal ossification, 163–64

metatarsals, 99

metopic suture, 150

Mias, 11

mice: nucleotide sequence of, 270; tooth structure of, 298

Michelangelo, 1

micro-complement fixation (MC'F), 230–31

Miocene fossils, 27–31, 72–73, 77–78, 232–33; *see also* hominoid evolution

missing-link research. *See* hominid evolution; hominoid evolution

Mitani, John, 12

mitochondrial DNA, 268–72; variability in, 299

Mivart, St. George, 127, 129–30, 131–32, 133, 134, 140, 141

M'Leod: description of orang-utan, 5–6, 7, 8

molar cingula, 203–4

molar enamel. *See* molar structure

molar structure: and anthropoid evolution, 149–50; catarrhine, 153; cingula, 293–4; cloning, 298; deciduous teeth, 299; and diet, 305; and evolutionary theory/research, 29–30, 38–39, 40–41, 74, 76, 78, 284; of gorillas and chimpanzees, 175; hominid, 118–19, 178, 284, 299; hominoid, 118–19, 165–66, 175, 178, 200–203, 283–89; of mammals, 131; of marsupials, 141; of New World monkeys, 38; of Old World monkeys, 128; of orang-utan, 118–19, 200, 284, 299; pattern 1 enamel, 284–88; pattern 3 enamel, 284–88; of *Ramapithecus*, 175; of *Sivapithecus*, 284–85, 287–88, 299; thick/fast enamel, 284–88; thick/slowed enamel, 284–88; third molar, 202–3, 298

molecular anthropology: origins of, 226; *see also* molecular research

molecular research, 55, 70–71, 81–82, 181, 226–52, 253–82, 291–92; by Benveniste/Todaro, 260–61; by Britten/Kohne, 258–59, 291; by Brown et al., 268–71, 272; conflicts with morphologists and paleontologists, 226–27, 234–37, 272, 277; dependence on experimental design, 291; disagreements among systematists, 291; DNA hybridization, 253, 256–65, 272, 291–93, 295; and genome significance, 295; by Goodman et al., 226–29, 231–33, 235, 237–40; by King/Wilson, 253–54, 291, 294; by Lowenstein et al., 277–81; by Romero-Herrera et al., 248–51; by Sarich/Wilson, 229–34; by Sarich/Cronin, 239–40; by Sibley/Ahlquist, 262–64; taxonomic assumptions of researchers, 272–73; by Zuckerkandl/Pauling, 241–42, 244

molecular structure: prepatterning vs. variability in, 298–99

molecular systematists. *See* molecular research

molecules: fetal, 222–23

Molleson, Theya, 40

monkeys. *See* marmosets; New World monkeys; Old World monkeys; primate(s)

monogamy: in gibbons and siamangs, 157; sexual dimorphism
and, 301
morphology: conflicts between
morphologists and molecular systematists, 226–27, 234–37, 272,
277, 296; and genomic dissimilarity, 294–96; lack of unanimity of theoretical and methodological principles, 291, 294–
95; and prepatterning vs. variability in anatomical systems,
298–99; and taxonomy, 100–
101
moths: establishing taxa for, 108–9
mountain gorillas, 2; *see also* gorillas
mouse lemurs: clawlike "nail" or
grooming claw of, 147; taxonomy of, 238; tooth comb of, 148
mRNA, 226, 257
muscular anatomy: and hominoid
relatedness, 289–90
myoglobin, 247–51, 266

Nadler, Ronald, 211–12
nails: in definition of anthropoids
and primates, 130, 132, 133,
141, 147, 151; prosimian, 147–
48
Napier, John R., 52
Nasal structure: "African" vs.
"Asian" nasal cavity pattern,
190–94; cartilage, 185; catarrhine, 152–53; and evolutionary
theory, 184–85, 186, 190–94;
and hominoid relatedness, 71–
72; of New World monkeys,
152, 167; sinuses in hominoids,
182–84, 186–87, 195
Nature, 217
Neanderthals: and evolutionary
theory, 31–32, 33, 34–35, 36,
38, 86; single incisive foramen
in, 177
neoteny, 162–63

New World monkeys (Platyrrhini),
88; anthropoid characteristics of,
151; classification of, 127, 128,
152; fossils, 143; as "higher"
primates, 127–28; and hominoid
relatedness, 225–26; molar
structure of, 38; molecular research on, 226, 230; as most
primitive of anthropoids, 128;
myoglobin research on, 249;
nasal structure of, 152, 167
nipples: hominoid, 210; *see also*
mammary glands
nostrils. *See* nasal structure
nucleotide(s), 240–42, 251–52,
256–57, 266–68, 297
nucleotide replacement (NR), 246,
251–52
nucleotide sequences, 242, 256–59,
260–61, 291, 293, 297
Nutcracker Boy, 50–52
Nuttall, George H. F., 71, 220–21,
223–26, 227, 273

Oakley, Kenneth, 40
obligate traits, 300–301, 302
Olduvai Gorge: research at, 48,
49–52, 53
Old World monkeys (Cercopithecidae), 38, 43, 88, 91, 103, 128;
anthropoid characteristics of,
151; bilophodonty in, 128, 159;
brow ridge development in, 195;
canines of, 166–67; catarrhinism
in, 152–53; classification of, 43,
127, 128; fossils, 142–43; as
"higher" primates, 127, 128;
and hominoid relatedness, 151–
53, 225–26; ischial callosities in,
209–10; migration from African
Plate, 304; mitochondrial DNA
of, 269; molar structure of, 38,
128; molecular research on, 226,
229, 230, 231, 234, 273; myoglobin research on, 249–50;
nasal structure of, 152–53, 167;

ossification in, 161–63; palatine fenestrae of, 168, 171; sacral segments in, 159–60; tooth structure of, 38, 128, 159, 166–67

Oligopithecus, 143

omnivores: hominids as, 48

Omo: evolutionary research at, 53

On the Origin of Species (Darwin), 22, 27, 30, 57, 59

ontogeny, 61–64

opossum: grasping foot of, 132; tooth structure of, 131

Orang Outang conglomerate, 16, 19–23, 27; *see also* orang-utans

orang-utans (*Pongo pygmaeus*): arborealism, 3, 5, 7, 11, 304–5; arms of, 5; "Asian pattern" nasal cavity, 190, 191, 193–94; basicranial region of, 206; body hair of, 115, 196, 284, 305; Bornean vs. Sumatran, 8–9; brain of, 66; brain asymmetry in, 214–15; canines of, 166–67, 201–2; carpal bones in, 188–89; chromosome banding pattern in, 274–76; common ancestor with *Homo,* 283–84; deciduous teeth in, 205; destructive behavior of, 11–12; diet of, 3–5, 12, 304–5; digits of, 5, 7, 65, 114; DNA hybridization research on, 264, 292; double-jointedness in, 91, 115; early research on, 2–25; epiphyseal ossification in, 198; estriol secretion in, 215–16; etymology of name, 16; evolution from *Sivapithecus,* 74, 78, 79, 83–85, 176–77, 179, 182, 190–94, 201, 284, 303; faces of, variations in, 8–9; facial hair of, 9; fibrinopeptide protein sequences, 292; foraging territorial needs, 12; foramen lacerum in, 207, 208; genetic distance from *Homo,* 255–56; genital swelling in, lack of, 212; genomic dissimilarity, 295–96; genomic novelty of, 292–93; gestation period of, 210–11; hair length of, 115, 196, 284, 305; handedness in, 214–15; hands and feet of, 5–6, 7, 114–15; hemoglobin of, 247; hip joints of, 6, 292; hominid features and relatedness, 56, 57, 67, 84–85, 113–15, 178–79, 181–218, 283–306; and hominoid relatedness, 114–20, 178–79, 283–306; hominoid uniqueness of, 114–15, 120, 292–93; as *Homo sylvestris,* 19–20; "humanness" attributed to, in early research, 16–17; ischial callosities, lack of, 210; legs of, 5–6; locomotion of, 7, 11; male territorial calls, 13; mammary glands in, 210; mandible structure and molar enamel, 38–39, 40, 41; mating and courtship behavior, 13–15, 212, 301–3; migration from African Plate, 304; as missing link in great chain of being, 19–23; molar structure of, 38–41, 118–19, 175–76, 178, 200, 286–88, 299; molecular research on, 228–29, 234, 247, 249–51, 254, 273, 274; myoglobin research on, 249–51; nasal structure of, 167, 190–94; nerves in incisive foramen, 293; and 19th century evolutionary research, 23–25; nucleotide mismatch with *Homo,* 261–62; palatine fenestrae in, 169, 171; palatine ridges in, 187; patterns 1 and 3 enamel, 286–88; and Piltdown Man, 278, 279–81; polygyny among, 301; and *Ramapithecus,* 55–56, 182; scapula of, 197–98; serum-antiserum testing, 279–80; sexual dimorphism in, 19, 301; single incisive foramen in,

orang-utans *(cont.)*
174, 178, 179, 193, 293; sinuses
in, 183–84; size of, 3, 10; socia-
bility of, 12–13; strength of, 11;
supraorbital region of, 195–96;
talus of, 199; third molar in,
202–3; throat sacs of, 9, 115,
292; toes of, 65; tooth structure
of, 118–19, 166–67, 175–76,
178, 200–202, 292, 299; trunk
structure of, 153; type C viral
gene in, 260; weight of, 3
orders: in taxonomy, 101
Osborn, Jeff, 298
ossification: in hominoids, 161–64,
173, 198, 284, 299–300
otters: tool use by, 96
Owen, Sir Richard, 59, 131
ox: myoglobin research on, 248–49

paedomorphism, 162
Palaechthon, 146, 147
palatine fenestrae: hominoid, 168–
69, 171, 173–74, 177, 178,
189–93, 205
palatine ridges: and hominoid relatedness, 186–87
Paleopithecus sivalensis, 73
Paleosimia rugosidens, 74
Pan. See chimpanzees
parallelisms: and evolutionary
theory, 95
Paranthropus robustus, 46–47, 49
parsimonious phylogeny, 245–49
patas monkeys, 103
pattern 1 enamel, 285–88
pattern 3 enamel, 285–88
Pauling, Linus, 241–42, 244, 264,
273
Peacock, Thomas Love, 21–22
Peking Man *(Sinanthropus pekinensis),* 41–42
pelvic structure: of primates and
hominoids, 153–54, 159, 165
penguin: myoglobin research on,
251

penis: of mammals, 132, 133
Pennant, Thomas, 123
petrosal bone: in hominoids, 206–7
phalangeal joints: and knuckle
walking, 94–95
phylogeny. *See* hominid evolution;
hominoid evolution; primate evolution
pig: tooth structure of, 131
Pilbeam, David, 56, 77–78, 84,
102, 175, 177, 190, 191, 200–
201, 217
Pilgrim, Guy, 73–74, 75, 77
Piltdown Man *(Eoanthropus dawsoni),* 36–40, 70, 278–81
Pithecanthropus erectus, 35–38,
39, 41, 44, 69
Pithecus antiquus, 28
placenta: anthropoid, 150–51; and
evolutionary research, 64; and
fetal development, 222–23;
hemochorial placentation, 150–
51, 233
Platyrrhini, 152; *see also* New
World monkeys
plesiadapiforms (Plesiadapiformes),
138, 145–47
Plesiadapis, 145–47
Plesianthropus transvaalensis, 45–46
Pliopithecus antiquus, 28–29, 78
Pollicata, 125
polyandry: anthropoid, 157, 301
polygyny: anthropoid, 157, 301;
shared trait in *Homo* and orang-
utans, 301
Pongidae. *See* anthropoids
Pongo, 17
Pongo pygmaeus. See orang-utans
porpoise: myoglobin research on,
248
postorbital bars, 132–33, 141,
145–46, 147, 149, 151
postorbital closure, 133, 149,
151
potto: myoglobin research on, 251;
taxonomy of, 237

Prakash, Om, 273–74
pregnancy: of common orang-utan/ hominid ancestor, 305; and es- triol secretion, 215–16; in homi- noids, 210–11, 233, 284, 300; and molar enamel, 286
premaxilla: hominoid, 168, 190– 91, 193
premolar structure: and evolution- ary research, 74, 76, 78; and hominoid relatedness, 299; of mammals, 131; of Old World monkeys, 128; *see also* molar structure
primate(s) (Primates): brain, in clas- sification of, 130–31; brain asymmetry in, 214–16; catar- rhinism, 152–53; crucial features of, 129, 141–42; deciduous teeth in, 204–5; defining, 121–55; der- ivation from insectivores, 137, 138–39; digits of, 130, 141; Flower's classification of, 125– 29; fossils of, 142–49 (*see also* primate evolution); frontal si- nuses in, 182–84, 186; gestation period of, 210–11; hands of, 122, 123, 132, 141, 147; hierar- chical ordering of, 15–16, 127– 29; "higher," 127–29; hip joints of, 6; Illiger's classification of, 125; Le Gros Clark on, 134–35, 137, 138–40; in Linnaean classi- fication, 122–23, 125–26, 129– 30, 141; "lower," 127; mam- mary glands of, 122; mating habits of, 13–15; in Mivart's classification, 129–30, 131–32, 133, 134, 140, 141; molar struc- ture of, 38; nails of, 130, 132, 133, 141, 147; nasal structure of, 151–53; 19th century re- search on, 23–25; pelvis of, 153–54; and plesiadapiforms, 138; Quadrumanes vs. Bimanes, 123, 125; shoulder joints of, 6;

Simpson on, 134–35, 137, 138; systematists on, 121–34; taxo- nomic boundaries of order, 121– 55; teeth of, 122, 123, 129–30, 131, 141–42, 146–47; and tree shrews, evolutionary link, 139– 40; trunk structure of, 153–54; wrist structure of, 188–89; *see also* anthropoids; hominoid(s); New World monkeys; Old World monkeys; prosimians
primate evolution, 26–56, 137–55; *see also* hominid evolution; hominoid evolution
primatologists: conflicts among, 134–37
Proconsul, 29, 75, 76, 78–79, 143, 200, 203, 232
prosimians (Prosimii), 125, 127, 147–48; and anthropoids, evolu- tionary relationship, 147; carpal bones in, 188; Eocene fossils of, 142–43; hands and feet of, 141; molecular research on, 231, 237–40, 266–67; morphological vs. molecular approach to, 237– 40; myoglobin research on, 249– 51; nasal structure of, 167; pala- tine fenestrae in, 168–69; post- orbital bars in, 133, 141; taxonomy of, 237–40; tooth comb of, 148–49; *see also* bush- babies; lemur(s); lorises; mouse lemurs; tarsiers
proteins, 240–41, 266; and DNA, 256; variability in, 299; *see also* albumin
pygmy chimpanzees, 2, 18; anti- serum testing with, 279–81; hemoglobin of, 247; molecular research on, 255; *see also* chim- panzees

Quadrumanes (Quadrumana), 123, 125
Quimpanzes, 18

Raikow, Robert, 263
Ramapithecidae, 81, 102
Ramapithecinae, 102
Ramapithecus, 55–56, 73–83, 102,
174–75, 187; canines of, 97;
and hominid evolution, 73–83,
174–77, 232, 235–36, 247;
Miocene dating of, 232; molar
structure of, 284; molecular systematists on, 82; relation to *Sivapithecus*, 83, 176, 182, 232;
taxonomic classification of, 102;
tooth structure of, 97, 175–76,
284
R. brevirostris, 75
reciprocity: test of, 230–31
regulatory genes, 253–54
reproductive behavior. *See* mating
behavior
Rhesus monkeys, 91; androsterone
in, 164; antiserum testing with,
279–80; molecular research on,
230
Rh factor, 221–23
ribonucleic acid (RNA), 226, 240–
41
rib structure: hominoid, 154, 159
Rijksen, Herman, 12
RNA (ribonucleic acid), 226, 240–
41
"rock badger," 100
rock hyrax, 100
rodents (Rodentia): cerebrum of,
130; evolution of, 80
Rodman, Peter, 12, 305
Romero-Herrera, A. E., 248–51,
265
ruminating, 99

sacrum: hominoid, 159–60
Saint-Hilaire, Isidore Geoffrey, 127
Sarich, Vincent M., 55–56, 82, 85,
229–32, 233–40, 246, 266, 270,
271, 272, 273, 296
scapulae: hominoid, 90, 154–55,
159, 165, 196–98, 284

Schultz, Adolph, 66–68, 70, 117,
158, 159, 160, 161, 183–84,
185–86, 189, 194–96, 208,
209–11
Schürmann, Chris, 12
seals: myoglobin research on, 248
serine, 250
serum. *See* blood serum research
serum-antiserum reaction, 223–26,
127, 229, 230–31, 279–81
serum test of reciprocity, 230–31
sex chromosomes: hominoid, 275
sex hormones: hominoid, 164–65
sexual behavior. *See* mating behavior
sexual dimorphism: and canine
size, 166; and diet, 300; and
evolutionary research, 52–53; in
Homo, 301; and mating system,
157; and monogamy, 301; in
orang-utan/hominid common
ancestor, 304
Shea, Brian, 283
sheep, 99; myoglobin research on,
248
Shideler, Susan, 165, 215–16
shoulder structure: and brachiation,
156; hominoid, 90, 154–55,
165, 196–98; primate, 6; *see
also* scapulae
shrews, 137; and *Plesiadapis,* 146
siamangs (Hylobatidae): anatomical
characteristics of, 157; as brachiators, 156–58; brow ridge in,
195–96; canines of, 166–67;
distribution of, 157; habitat of,
157; as hominoids, 158–59;
mating habits of, 157; molar
structure of, 202; molecular research on, 260; myoglobin research on, 249; nasal structure
of, 167; ossification in, 163;
sacral segments of, 159; trunk
structure of, 153
Sibley, Charles, 262–64, 265, 271,
272, 295

sifakas: taxonomy of, 237–38; *see also* lemur(s)

sight: and evolution of primates, 145–46, 149

sign language communication, 215

Simia, 122

Simiidae, 129

Simons, Elwyn, 75–76, 77, 84, 102, 107, 142, 143, 174–75, 235

Simpson, George Gaylord, 102, 134–35, 137, 138, 139, 145

Sinanthropus pekinensis (Peking Man), 41–42

single incisive foramen. *See* incisive foramen, single

sinuses, frontal: hominoid, 71–72, 182–84, 186–87, 195

sirenians: cerebrum of, 130

Sivapithecus, 73, 74, 78, 79–81, 83–85, 143, 175, 217; and "Asian pattern" nasal cavity, 190, 191, 193–94; canines of, 201–2; and evolution of orangutan, 74, 78, 79, 83–85, 176–77, 179, 182, 190–94, 201, 284, 303; molar structure of, 203–4, 284–85, 287–88, 299; relation to *Ramapithecus*, 83, 176, 182, 232; single incisive foramen in, 193

S. indicus, 74

S. sivalensis, 73

Siwalik Hills, India: fossil deposits at, 73, 75, 175

skull structure: anthropoid, 149–50; basicranial region, 205–7; brow ridge development, 195–96; catarrhine, 153; and evolutionary theory, 31, 34–35, 36–39, 42–43, 51–55; and hominoid relatedness, 185–86; postorbital bar, 132–33, 141, 145–46, 147, 149, 151; and sinuses, 71–72, 182–84, 187, 195; soft tissue determination, 298

sloths (*Bradypus*), 122, 123, 130; cerebrum of, 130; in Linnaean classification, 122, 123; mammary glands of, 131

slow lorises, 207–8; myoglobin research on, 251; *see also* lorises

smell: sense of, and primate evolution, 145, 149, 151

Smith, G. Elliot, 70, 113

Smith Woodward, Sir Arthur, 37–39, 40

snouts. *See* nasal structure

southeast Asia: evolutionary research in, 2–3, 33–37

species: in taxonomy, 101

sportive lemurs, 238; myoglobin research on, 249; *see also* lemur(s)

Spy Neanderthals, 32–33

squirrel monkeys: myoglobin research on, 249

stem cells, 298; of molars, 299

Sterkfontein cave site, 45–46

steroids, 164

structural genes, 253

Sumatra, 16

Sumatran orang-utans, 16; description of, 9–10; face of, 8–9; types of, 9–10; *see also* orangutans

superfamily: in taxonomy, 101

supraorbital thickening, 195–96

Swarts, J. Douglas, 205

Sylvian sulcus, 213–14, 300

Systema Naturae (Linnaeus), 122, 125, 127, 129

talus: hominoid, 199

tarsiers, 88–90, 125, 126, 232; fossils of, 142; grooming claw of, 147, 148, 149; hemoglobin of, 245–46, 248, 266; and hominid evolution, 88–90; lack of tooth comb, 148–49; as link between lower and higher primates, 127, 128, 148; molecular research on, 238–40, 244–46, 248; postorbi-

tarsiers (*cont.*)
     tal bar in, 133; taxonomic classi-
     fication of, 127, 238–40, 244–
     46
Tarsiidae, 127
Tashian, Richard, 237
Tasmanians, 71–72
Tasmanian wolf, 278
Tattersall, Ian, 142, 208, 238
Taung site fossil, 42–43, 44, 45,
     46, 49
taxonomy: arbitrariness of, 102–3;
     biblical records of, 98–99; for
     communication, 106; conflicts
     between molecular systematists
     and morphologists/paleontolo-
     gists, 226–27; establishing taxa,
     108–12; function of, 100–104;
     lack of unanimity in theoretical
     and morphological principles,
     291; and molecular research,
     219–82; of primates, 121–34;
     problems of, 103; uniqueness of
     species, 111
T-cell research, 273–74
Teilhard de Chardin, Pierre, 40
Tekkaya, I., 83
Telliamed (de Maillet), 6, 17
Templeton, Alan, 265
temporal muscles: in mammals,
     132–33
teres ligament, 115
testes: of carnivores, 131
third molar: hominoid, 202–3; of
     mice, 298; *see also* molar struc-
     ture
thoracic structure: hominoid,
     154
thumbs: opposable, in primates,
     65, 132, 141
Tobias, Phillip V., 52
Todaro, George, 260–61
toe: big, of mammals, 132, 133,
     141; of primates, 132, 141
tool making: and evolutionary
     theory, 96
tooth comb: prosimian, 148–49

tooth structure: and anthropoid
     evolution, 149, 156; catarrhine,
     153; cloning in, 298; deciduous
     ("milk") teeth, 204–5; determi-
     nation of, 297–98; and evolu-
     tionary theory, 28–31, 38–39,
     40–42, 51, 54–56, 74, 76, 78,
     79–80, 95–97, 175–79, 199–
     200, 284; and hominoid related-
     ness, 199–205; of mice, 298; of
     Old World monkeys, 128; of
     orang-utan, 175–76, 200; plesia-
     dapiform, 145, 146; of primates,
     122, 123, 129–30, 131, 141–42,
     146–47, 165–66; prosimian,
     148; of *Ramapithecus*, 175–76;
     see also canines; molar structure;
     premolar structure; third molar
transferrin, 228, 229, 239
tree shrews (tupaioids), 131, 132;
     as insectivores, 137, 139; as
     missing link to primates, 139–
     40; molecular research on, 240;
     myoglobin research on, 251
trunk structure: hominoid, 153–54,
     156
tupaioids. *See* tree shrews
Turkana: evolutionary research at,
     52–53
Tuttle, Russell, 94
twins: identical, 219
two-handedness: of *Homo*, 123, 125
type C viral gene, 260
Tyson, Edward, 19–20

Uhlenhuth, Paul, 223–25, 227
ulna: hominoid, 155, 159, 173;
     ossification, 163, 164, 198
ungulates: cerebrum of, 130
uniform average rate (UAR) of ge-
     nomic change, 262–63
uterus: bicornuate, 150; simplex,
     150, 151

vertebral structure: hominoid, 159–
     60
Vespertillio, 122, 123; *see also* bats

vision. *See* eye; sight
vomeronasal organ, 171
von Baer, Karl Ernst, 62–63
Vrba, Elizabeth, 198

Wallace, Alfred Russel, 3, 10–12
Ward, Steve, 190, 200, 201–2, 293
Wasserman, E., 230
Watson, James Dewey, 226
Wayland, E. J., 28
Weidenreich, Franz, 42, 278
Weiner, John, 279
Weinert, H., 67–68, 183, 187–89
Werth, E., 69
whales, 100; cerebrum of, 130;
    myoglobin research on, 248
White, Tim, 177, 205–7
Wicker, Fred, 77
"Wild man" (Linnaeus), 20
Wilson, Allan C., 55–56, 82, 85,

229–32, 233–34, 237, 253–54,
    268, 270, 271, 272, 273, 291,
    294
Winkler, Susan, 290
"wisdom tooth," 202
wolf: Tasmanian, 278
woolly monkey: myoglobin re-
    search on, 249
wrist structure: and brachiation,
    156; hominoid, 155, 159, 173,
    188–89

Xenopithecus, 29

Yunis, Jorge, 273–74

zebra: myoglobin research on, 251
Zinjanthropus, 50–52, 97, 177
Zuckerkandl, Emile, 241–42, 244,
    264, 273